Sustainable Land Management

Strategies to Cope with the Marginalisation of Agriculture

Edited by

Floor Brouwer

Head, Research Unit on Management of Natural Resources, LEI Wageningen UR, the Netherlands

Teunis van Rheenen

Coordinator for Partnerships & Impact Assessment and Secretary to the Board of Trustees, International Food Policy Research Institute (IFPRI), USA

Shivcharn S. Dhillion

Independent Consultant of ENVIRO-DEV, Norway

Anna Martha Elgersma

Researcher at the Norwegian University of Life Sciences (UMB), Department of Ecology and Natural Resource Management, Norway

Edward Elgar
Cheltenham, UK • Northampton, MA, USA

Published by
Edward Elgar Publishing Limited
The Lypiatts
15 Lansdown Road
Cheltenham
Glos GL50 2JA
UK

Edward Elgar Publishing, Inc.
William Pratt House
9 Dewey Court
Northampton
Massachusetts 01060
USA

A catalogue record for this book
is available from the British Library

Library of Congress Control Number: 2008932862

ISBN 978 1 84542 902 7

Printed and bound in Great Britain by MPG Books Ltd, Bodmin, Cornwall

Contents

v

Contributors

Arnold Arnoldussen, Head of the Section Land Resources at the Norwegian Forest and Landscape Institute, Norway.

Eva Boucníková, PhD student at the Faculty of Biological Sciences at the University of South Bohemia, Czech Republic.

Bas Breman, Research Scholar at Alterra, Wageningen University and Research Centre, The Netherlands.

Floor Brouwer, Head of the Research Unit on Management of Natural Resources, LEI, Wageningen University and Research Centre, The Netherlands.

Dominique Cairol, Researcher at the Agricultural and Environmental Engineering Research (Cemagref), France.

Patrick Caron, Scientific Director and Head of the Department Environnements et Sociétés, CIRAD, France.

Thomas Dax, Deputy Director of the Federal Institute for Mountainous and Less-Favoured Areas (Bundesanstalt fuer Bergbauernfragen, BABF), Austria.

Shivcharn S. Dhillion, Independent Senior Consultant of ENVIRO-DEV, an Environment and Development Consulting Firm, Tylldalen, Norway.

Michael Duffy, Professor of Economics and the Director of the Beginning Farmers Center, Iowa State University, USA.

Anna Martha Elgersma, Researcher at the Norwegian University of Life Sciences, Department of Ecology and Natural Resource Management, Norway.

Josef Fanta, Professor Emeritus, University of Amsterdam and Wageningen University, The Netherlands, and Visiting Professor (landscape ecology) in the Faculty of Biology, University of South Bohemia, Czech Republic.

Alajos Fehér, Principal Scientific Fellow and Visiting Professor in the University of Agricultural Sciences, Hungary.

Yoichi Matsuki, Professor in the Department of Nature Management and AgriFood Economics, Nippon Veterinary and Life Science University, Tokyo, Japan.

Miki Nagamatsu, Associate Professor in the Department of Nature Management and AgriFood Economics, Nippon Veterinary and Life Science University, Tokyo, Japan.

Teresa Pinto-Correia, Associate Professor in the Department of Landscape and Biophysical Planning and Co-ordinator of the Research Group Mediterranean Ecosystems and Landscapes, University of Evora, Portugal.

Teunis van Rheenen, Coordinator for Partnerships & Impact Assessment at the International Food Policy Research Institute (IFPRI), Washington, DC, USA.

Jordi Rosell, Professor in the Applied Economics Department, Universitat Autònoma de Barcelona, Spain.

Gábor Szabó, Professor in the Department of General and Agricultural Economics at the University of Debrecen, Hungary.

Marja-Liisa Tapio-Biström, Senior Adviser in the International Affairs Group, Ministry of Agriculture and Forestry, Finland.

Hilkka Vihinen, Professor of Rural Policy and Head of the Research Unit on Rural Policy at MTT Economic Research, Helsinki, Finland.

Lourdes Viladomiu, Professor in the Applied Economics Department, Universitat Autònoma de Barcelona, Spain.

Georg Wiesinger, Rural Sociologist and Senior Researcher at the Bundesanstalt für Bergbauernfragen (Federal Institute for Mountainous and Less-Favoured Areas), Austria.

Anna Zamora, Research Assistant at Universitat Autònoma de Barcelona, Spain.

František Zemek, Researcher at the Institute of Landscape Ecology, Czech Academy of Science, Czech Republic.

List of abbreviations

ASEAN	Association of South East Asian Nations
AWU	annual work unit
CAD	Contrat d'Agriculture Durable – Sustainable Agriculture Contract (French)
CAP	Common Agricultural Policy
CEE	Central and Eastern European (countries)
CORC	Concept Oriented Research Cluster
CRP	Conservation Reserve Program
CSP	Conservation Security Program
CSS	Cereal Steppes Scheme
CTE	Contrat Territorial d'Exploitation – Territorial Management Contract (French)
EAFRD	European Agricultural Fund for Rural Development
EAGGF	European Agricultural Guidance and Guarantee Fund
EQIP	Environmental Quality Improvement Program
ERS	Economic Research Service
ESU	Economic Size Unit
EU	European Union
FAO	Food and Agriculture Organization (of the United Nations)
GAEC	Good Agricultural and Environmental Condition
GMO	genetically modified organism
HNV	high nature value (farming system)
IMRI	Integrated Marginalisation Risk Index
IPCC	Intergovernmental Panel on Climate Change
JAS	Japan Agricultural Standard
LAG	local action group
LFA	Less Favoured Area
MF	multifunctionality
NGO	non-governmental organisation
OECD	Organisation for Economic Co-operation and Development
ROA	Roles of Agriculture (project)
SARD	Sustainable Agriculture and Rural Development
SARD-M	Sustainable Agriculture and Rural Development in Mountains
SD	sustainable development

SEO	Sociedad Española de Ornitología (Spain)
SEPFF	Protection Programme of Flora and Fauna in extensive agriculture areas (Spain)
SGM	standard gross margin
SMR	Statutory Management Requirement
UAA	utilised agricultural area
UNCED	UN Conference on Environment and Development
URAA	Uruguay Round Agreement on Agriculture
USDA	United States Department of Agriculture
WTO	World Trade Organization

Preface

This book presents and explores strategies to reverse the process of marginalisation of rural areas, looking at the economic viability of land, including the suitability for different land use options, giving consideration to the needs for environmental conservation. This volume takes the reader beyond the established knowledge that marginalisation of land through agricultural marginalisation is occurring. The complexities and driving forces governing marginalisation are not always the same across nations and regions due to climatic, geographic, economic, legislative and political status. This book illustrates the linkages between these issues and disentangles the complex relations at the national and regional levels. Illustrative cases looking at the national and regional trends across European nations, Japan and the USA are presented, specifically looking at strategies for coping with marginalisation. This volume has a strong 'lessons to be learnt' emphasis, to facilitate more sustainable land use in the future. The need to understand and cope with marginalisation processes arose at the same time as the development of the concept of multifunctionality, which also gained a vital place in the repertoire of coping strategies. Several chapters in this book also expound on the concept of multifunctionality at different stages in its development. The focus of the book is on marginalisation processes. A better understanding of such processes is essential to develop mitigating policies and actions that will slow down marginalisation.

The book is an effort by many contributors whose input is much appreciated. Many of the chapters arise from the results of a research project, Strengthening the Multifunctional Use of European Land: Coping with Marginalisation (EUROLAN). The project was funded by the European Commission under the RTD Programme Quality of Life and Management of Living Resources (Contract QLK5-CT-2002-02346), the support for which is gratefully acknowledged. In addition there are chapters solicited to enrich the range of perspectives in the volume. We appreciate the support given by Miss Felicity Plester (Edward Elgar Publishing).

Floor Brouwer, Teunis van Rheenen, Shivcharn S. Dhillion and Anna Martha Elgersma
June 2008

1. Introduction

Floor Brouwer and Teunis van Rheenen

Depopulation, intensification and extensification in agriculture and forestry and the establishment of protected areas for nature and landscape conservation can be considered as the main changes in land use in rural areas in the second half of the twentieth century. There has been a trend for many peripheral areas to become more depopulated than they are already, and for agricultural land to become more intensively or extensively used, or abandoned. The trends of depopulation, closing down of farms and land abandonment may indicate vulnerability to marginalisation of agriculture and of other land uses.

Marginalisation has to be considered as being caused by a complex of several external and internal driving forces. Globalisation, liberalisation of trade, European and national agricultural policies are mentioned as external forces for marginalisation, whereas opting for another lifestyle, unemployment, lack of an appropriate job, low income, lack of a successor in agriculture and a lack of capital for investment can be considered as internal forces.

Throughout its long history, agriculture has structured the cultural landscape and 'man-made' ecosystems more than any other human activity in many countries. Intensification and extensification in the past half-century has caused marginalisation of many traditional farms and environmental degradation due to the change from agricultural systems adapted to the local environmental conditions to systems not adapted to these conditions.

Large-scale farming has replaced small-scale farming even in areas where large-scale farming is hard to realise and farming has become specialised. In general, agriculture has shifted from a multifunctional culture or lifestyle towards an economic sector with specialised production systems. In many areas, often areas marginal for agriculture, these developments have led to the exodus of farming.

Raising income is one goal for coping with marginalisation of agriculture. The shift from small-scale farming to large-scale farming is an important way to increase income. Specialisation of the mode of production, intensification and more efficient management are other ways. Also, pluriactivity and having an off-farm job are ways to increase income. At present there is an increasing focus on the contribution of the multifunctionality of agriculture, and thus the integration of the functions of agriculture, to increase the viability of agriculture and its contribution to the viability of the rural environment.

In order to create a common understanding of marginalisation and multifunctionality, definitions have been developed that reflect their interdisciplinary nature:

- Marginalisation of land use is a process, driven by a combination of social, economic, political and environmental factors by which the use of land for the main land-dependent activities (agriculture, forestry, housing, tourism, local mining) ceases to be viable under an existing socio-economic structure.
- Multifunctionality of land use refers to the functions sustained by the land resources beyond their primary production functions (non-commodity products) such as, among others, environment preservation, cultural heritage, nature conservation and employment.
- Marginalisation of agriculture is a process driven by a combination of social, economic, political and environmental factors by which in certain areas farming ceases to be viable under an existing land use and socio-economic structure and no other agricultural options are available, so the process ends with land abandonment.
- Multifunctionality of agriculture is a socially constructed concept, which recognises that agriculture beyond its primary role of producing food and fibre also provides other functions such as the viability of rural areas, food security, the cultural heritage and environmental benefits such as the agricultural landscape, biodiversity and land conservation.

Marginalisation can be understood as a multidimensional process encompassing not only land abandonment, environmental degradation and economic decay, but also social and cultural patterns. Agricultural marginalisation can hardly be unravelled from the overall marginalisation in rural regions, but it implies economic, environmental and social marginalisation particularly in regions where agriculture is still a major economic sector. Agricultural policies might be insufficient to cope with the marginalisation of land. More integrative efforts might be needed with regional, environmental, socio-cultural and economic development in the attempt to combat marginalisation and land abandonment in rural areas.

Multifunctionality has long been viewed as a relevant issue and an increasingly important topic in agricultural and rural development policies. There is a prolific discussion on positively valued non-commodity outputs and also on negative externalities. The discussion on agricultural land use change is less intensive since there is no large-scale land abandonment. Since this process of land use change is going on rather slowly, it is not recognised as a major problem by the public. Yet, marginalisation is more important when addressing rural development, social and economic problems like unemployment, rural poverty, migration and the over-ageing of the population, and the degradation of amenities, infrastructure and services in remote areas.

Agricultural initiatives and pluriactivity are very important for combating marginalisation and land abandonment. Approximately 60 per cent of all farms in Austria are part-time farms integrating agricultural and non-agricultural activities to enhance their income. Pluriactivity plays an important role in sustaining cultivation, settlement, landscapes and cultural viability particularly in high-mountain and less-favoured regions. Moreover agricultural diversification and para-agricultural activities like food processing, direct marketing, processing of non-food products, cooperative farming activities, the provision of social services (schools, kindergartens, care places for elderly and disabled persons, drug and alcohol addicts, the homeless and so on) also contribute to the farmers' income.

KEY OBJECTIVES AND ORGANISATION OF THE BOOK

Marginalisation is observed in large parts of the world, and socio-economic and demographic factors are key driving forces. Strategies to enhance and strengthen the viability of agriculture and, more broadly speaking, the viability of the countryside are emphasised in this book. Inevitably the focus will shift to the multifunctionality in the context of providing support to farmers as guardians of the land, and changing societal demands and the reforms of the Common Agricultural Policy (CAP). Multifunctionality is a possible route to cope with marginalisation and to improve the viability of agriculture. Viability of agriculture will not be achieved without a thorough understanding of marginalisation and multifunctionality in a range of conditions.

The above-mentioned developments in agriculture and in rural areas are the basis of this volume, and its main objectives are:

- To study vulnerability to marginalisation and strategies to cope with the marginalisation of agriculture across rural areas in a range of conditions.
- To improve the understanding of land use types that might be most vulnerable to marginalisation.

- To explore the role of agricultural policies and other areas of action for coping with the process of marginalisation.
- To develop strategies for appropriate land use based on multifunctionality given the vulnerability for marginalisation.

The first part of the book primarily offers an in-depth overview of concepts of marginalisation and multifunctionality across Europe, and an empirical identification of hot spots of marginalisation according to some selected risks and driving forces.

Chapter 2 offers a reflection on the concept of marginalisation. Teresa Pinto-Correia and Bas Breman present an overview and discussion of various concepts of marginalisation. A distinction is made between marginalisation of land, of agriculture and of rural areas, and linkages are established with the occurrence of land abandonment. The concept has evolved since the late 1980s and the period leading to the CAP reform of 1992. Examples from Portugal (a country where a relatively large proportion of the land is considered marginal) are compared with other parts of Europe. A decline in food production is compared to the increase (maintenance) of other functions of land, and the challenges of multifunctionality are clarified. The authors conclude that marginalisation is largely seen as a process of relative change, driven by economic and social factors. Clarifications provided in the chapter are vital to support decision-making processes concerning the future of rural areas.

Chapter 3 reviews the different meanings of multifunctionality, and the relevance of the concept to design strategies to cope with marginalisation. Patrick Caron and Dominique Cairol clarify the different meanings of multifunctionality, based on a comparative analysis of case studies throughout Europe. Several Concept Oriented Research Clusters (CORCs) are identified. The authors also clarify the long-term viability of the agricultural sector to compete, and revitalisation of rural areas is looked at as a strategy to cope with marginalisation processes. Solutions are suggested to strengthen the interaction between agriculture and the sustainability of rural and urban areas. A territorial approach is used to understand and cope with marginalisation.

Chapter 4 examines the issues of marginalisation as they relate to circumstances in the United States. Michael Duffy identifies the main demographic and economic changes taking place in rural communities. The author illustrates the bimodal nature of farms, with the very small farms representing the majority of holdings and a small number of holdings account for the largest part of production value. The ageing of the population is another significant change occurring in agricultural and non-farming communities. Coupled with a decrease in the counties dependent upon agriculture for their income, there also has been a decrease in the economic viability of many rural counties. There is very little farmland that is abandoned without another use, but a considerable amount of farmland is under pressure to change its use. He argues that multifunctionality might strengthen the link between agriculture and the rural communities, but it remains to be seen whether such an approach will help alleviate the problems we observe in agriculture and rural communities.

Part II of the volume explores working strategies to strengthen the viability of agriculture. In-depth analyses from a range of case studies are presented to improve the understanding of which land use strategies have proven to be successful to cope with marginalisation of agriculture. Attention is given to thematic changes that refer to broad trends, instead of directly referring to single national case studies.

Chapter 5 explores how the difference between the north and the rest of Europe, namely a colder climate, sparse population and large distances between villages, has had an impact on the rural areas. These conditions are serious handicaps for agriculture and rural development. Shivcharn Dhillion and his co-authors analyse several indicators and look at marginalisation coping strategies. The authors show how the availability of off-farm jobs is crucial for combating marginalisation. However, at the same time finding an off-farm job may be a problem in sparsely populated areas with long distances to centres of population. The specialisation of areas and specialty product development is a viable aspect for future development.

Marginalisation processes of rural economies in the Czech Republic and Hungary are examined in Chapter 6. Alajos Fehér and his colleagues identify the transition processes that took place in these countries following the collapse of the Communist system and they evaluate their consequences for the rural economy. They show how the viability of rural economies was threatened in areas with marginalisation of land. The period of societal change from centrally-planned to market-driven economies has been characterised by social uncertainty, lack of appropriate tools to manage change and uncertainty about ecological impacts. Several processes are identified to strengthen the development and diversification of regional economies (including agriculture and specific services such as tourism, protection of the environment and maintenance of biodiversity).

Marginalisation processes in mountain areas are central to Chapter 7. Thomas Dax and Georg Wiesinger provide empirical evidence on the harmful effects of abandonment of farm management in sensible ecological conditions, not only for the agricultural sector, but also for the regional economy and the viability of social structures. Large areas in the mountain region have seen a trend of depopulation. Quality of agricultural produce and the provision of region-specific products represent a major asset of mountain farming, especially in areas where rural amenities like landscape and nature are appreciated for tourism. Also, the integration of off-farm labour, pluriactivity and regional policies is a prerequisite for the long-term provision of social demands.

Chapter 8 examines marginalisation in Spanish dry areas. Jordi Rosell and his colleagues explore ecosystems in Southern European regions that are fragile, with high risks of soil erosion, forest fires and desertification. Population is also very unevenly distributed, with rural areas facing severe depopulation. Demographic and economic factors are key drivers of the vulnerability to marginalisation and the risk of land abandonment. The authors also indicate vulnerability to land abandonment due to the decoupling of farm support measures from production, especially on land with lower productivity and high levels of fallow land. For that reason, measures for Good Agricultural and Environmental Conditions (GAECs) have been introduced with the 2003 reform of the CAP. In order to cope with the marginalisation of agricultural land, the authors recommend economic diversification in rural areas with high potential for tourism, with multifunctionality to be a basis to reinforce other activities besides farming.

Chapter 9 provides an attempt to quantify marginalisation of agricultural land. Teunis van Rheenen and Floor Brouwer offer an indicator-based approach to identify areas at the highest risk of being marginalised. A methodology is proposed to quantify a so-called Integrated Marginalisation Risk Index (IMRI). The IMRI proposed is a combination of indicators on agricultural income as a percentage of non-agricultural income, share of agricultural holders that are older than 55 years, population density (number of inhabitants per square kilometre), and erosion (annual soil loss per hectare). The authors conclude that marginalisation might be invisible in many parts of Europe until it is too late to take action, and it will be become more difficult to conduct damage control.

Part III of the volume explores sustainable land management practices. This part of the work is forward-looking, offering a basis to agriculture and policy for strategies and best practices that are suitable to contribute to sustainable land management practices.

Chapter 10 is about high nature value supply chains in Japan. Yoichi Matsuki and Miki Nagamatsu examine recent developments of the agri-food system, and how co-operation between producers' and consumers' organisations has been strengthened to promote organic production, controlling the use of pesticides and fertilizers. Consumers move from food safety and food security to a new stage of nature protection and conservation of the environment. Agri-food supply systems are developed that revitalise rural communities and supply environmentally friendly food.

Chapter 11 examines how social capital can be crucial to gain a better understanding of marginalisation. Georg Wiesinger and his co-authors describe how policy measures and social capital interact with marginalisation. Case studies are undertaken in three EU member states (Austria, Finland and Hungary) with different socio-economic and demographic features and wide differences in social cohesion and the innovative ability of society. The existing social capital at local level may play a decisive role in facilitating the utilisation of natural and human resources through social networks, trust and civicness.

Chapter 12 strengthens the understanding of the quality of land in agricultural land use systems. Anna Martha Elgersma and her co-authors argue that the physical and biological aspects of land and its soil have been given insufficient priority in the designation of land use. As a result, such inadequate understanding of land quality aspects can lead to serious burdens on society. A suitability flow-chart and a land classification system are proposed to support sustainable planning of land use at the field, farm, local, regional and higher scales. They would be important building blocks for developing policy strategies. The authors conclude that changes in the quality of land use have their basis grounded in intensification and scale. Suitability flow-charts and land classification systems may help to facilitate the inclusion of properties related to productivity, biodiversity and make-up. It is seen as important that such properties are included in local and regional management schemes and policy measures.

Chapter 13 reflects on changing land management practices and explores emerging perspectives. Floor Brouwer and his co-authors reiterate the desire for tools that are generic and sufficiently flexible to accommodate local and regional characteristics. They argue that pluriactivity will facilitate multifunctional land use, and the integration of the farming communities into the rest of society is seen as a vital strategy to cope with marginalisation, and to broaden the economic basis for viable land management practices.

PART I

UNDERSTANDING MARGINALISATION
AND MULTIFUNCTIONALITY

2. Understanding marginalisation in the periphery of Europe: a multidimensional process

Teresa Pinto-Correia and Bas Breman

The concept of marginalisation, related both to problems of viability of agricultural activity in less-favourable areas, and to the decay of rural communities, was put to the debate in the European context in the late 1980s and beginning of the 1990s. The concern of policy-makers and the interests of research at that time were related to the first reform of the Common Agricultural Policy (CAP) and uncertainties as to the impact of the changes in the price and support mechanisms for agricultural production. Most scenarios predicted high rates of abandonment of agricultural land, even in countries with a modern and competitive agricultural sector, since under the foreseen conditions it would not be economically rational to continue farming in land with less-favourable conditions or locations (Bethe and Bolsius, 1995; Cabanel and Ambroise, 1990; Reenberg and Pinto-Correia, 1993).

These scenarios were developed mainly for north-western countries in the European Union (EU), and they related mainly to the marginalisation of land. Later, it has been shown that in the less-favourable areas, land use has been extensified, and land has sometimes been abandoned, but that this was part of an overall rationalisation process where intensification was still the dominant process at the regional level. And at the same time, marginalisation of land did not reach the levels that many scenarios foresaw. Furthermore, the extensive use of parts of the farm unit could be supported by external funds due to the associated environmental and landscape benefits. This resulted in another discourse, where marginalisation of land in relation to intensive farming uses has been seen as positive. This valorisation is due primarily to its ecological benefits and its contribution to landscape diversification (Benjamin et al., 2005: 630; Brandt and Primdahl, 1994). Thus, even if the land is not used for intensive agricultural production, it may not be seen by land owners as marginalised, but rather as a different type of land with a different use and function.

In this way, most of the debate and research on marginalisation in north-western countries of Europe has progressively shifted focus, often to the

11

extensification of farming systems and the value and functions of non-intensive agricultural land, under the so-called high nature value (HNV) farming systems. After high visibility at the end of the 1980s and the beginning of the 1990s, published scientific work on land abandonment or marginalisation in rural areas became rarer later on.

Nevertheless, in the most peripheral areas of Europe, such as parts of the Mediterranean or Scandinavia, most of the mountain areas, and also in Eastern Europe, marginalisation processes have continued to take place, associated with different causes and with various types of consequences (Gomez-Limon and Lúcio, 1999: 167; Moreira et al., 2001: 558; Romero-Calcerrada and Perry, 2004: 225). Trends of marginalisation have been ongoing in such regions, both at a field level and on a wider scale, in different periods and with different intensities. From the start of the debate during the 1980s, significantly less research and public debate has been concentrated on such processes in these peripheral areas. Therefore they have only slowly, during the 1990s, been acknowledged in the European sphere as complex and multidimensional, since it is not only the land, but also the agricultural activity and the rural communities that can be seen as marginalising.

In the Mediterranean, the concept of marginalisation 'is mostly used in the context of a global marginalisation (concerning agriculture, but also economic and social factors) at a larger scale, at sub-regional or regional level, and also at national or European level' (Pinto-Correia, 1991: 99). By investigating the internal and external forces of marginalisation in the north of Portugal, Black (1992: 34) demonstrated that natural circumstances can never be seen as the only explanation of marginalisation processes. At least as important are the dynamics of a rural society, not least through its external links and contacts. The marginal characteristics from a natural perspective are often enforced by a lack of regional dynamism which is characteristic of peripheral regions, being the 'hard nucleus of rural marginality made up of those rural areas that are most disfavoured, both from a biophysical as well as from a socio-economic perspective' (Belo Moreira, 2001: 133).

But even if the combination of several types of socio-economic and biophysical characteristics reinforces marginalisation trends, a detailed analysis shows that marginal socio-economic conditions do not always go hand in hand with the marginalisation of land. Baptista (1995: 316) has made an important contribution, by describing how the destinies of productive agricultural activity and the population living on agricultural holdings have been decoupled: 'Agriculture no longer unites rural society with the whole of non-urban space. There now arises an issue of space, which is no longer part of agriculture and also no longer guarantees the vitality of rural society. The paths of agriculture, space and rural society are now dissociated.'

Furthermore, there is an issue of value. In many of the peripheral regions of Europe, contrary to what happens in the regions of intensive and specialised farming, the abandonment of farm use and thus the decay of the specific agrosystems is considered as a loss, both from an environmental and a cultural perspective. In the Mediterranean, over centuries, in response to the diversity of habitats, poor and stony soils, and climatic fluctuations, agriculture has developed mixed land use systems, mainly based on permanent crops (that is, olive groves, vineyards, orchards) and integrating open and wooded areas. The variety of crops is very high, leading to a diversified mosaic, and a valued heritage, high biodiversity, and generally highly multifunctional landscapes (Naveh, 1991: 545; Pinto-Correia and Vos, 2004: 136). Thus, slowing down or encountering marginalisation trends is often a policy concern in order to maintain valuable farming systems and the associated landscapes and rural communities.

The type of discourse associated with the use of the term 'marginalisation' is highly relevant in this context, as it contributes to the spontaneous and common understanding of the concept, and in consequence to the way in which the word is also used by the scientific community. There may then often be misunderstandings due to a lack of clarification.

Considering the above-mentioned various dimensions of the process of marginalisation, or better, the various processes that can be included under this broad umbrella, a renewed attention for the issue of marginalisation in Europe, its dimensions, drivers and consequences, demands an evaluation of the kind of processes occurring.

This chapter aims to contribute to a new and more comprehensive understanding of the question, through:

- a discussion and tentative definition of marginalisation and of relevant concepts that happen to be connected or used with similar meanings;
- an evaluation of the various dimensions of change taking place nowadays in the rural space, applied to Portugal, a country with dominant marginal characteristics in relation to European agriculture; and
- a tentative approach to relate the evaluation of marginalisation to the concept of multifunctionality of rural space, according to which agricultural activity and production are considered as one of many possible functions, leading to a new and more relative evaluation of marginalisation.

New insights are created, as the decline of agriculture may be compensated for by an increase in the importance of other functions, or these other functions may justify the maintenance of an otherwise threatened agriculture.

TOWARDS A CLARIFICATION OF CONCEPTS

Some terms commonly used in the scientific literature dealing with rural areas and agriculture are closely connected with marginalisation and warrant explanation, as their use may often be misleading. In everyday language, 'marginal' in spatial terms is associated with the periphery, as marginal areas or land are not central, from a geographic or economic or social perspective, or a combination of more than one of those. Furthermore, the term 'marginalisation' in relation to agriculture or agricultural use of the land is often used in the same way as 'abandonment'. In peripheric regions of Europe, as in the Mediterranean or Scandinavia or mountain areas, dealing with marginalisation is often also associated with depopulation, or social decay. 'Desertification' previously had mainly a physical meaning but nowadays is also used in relation to social processes, relating to depopulation.

Periphery

As opposed to the other concepts, 'periphery' does not relate to a process, but more to a condition, a characteristic. In its strict sense, periphery means the external contours, the surrounding zone of a core area. 'Periphery' is actually used mainly with reference to what is located on the fringe, on the border of what is central. At the European scale, the periphery is composed of the Mediterranean regions, eastern countries, Scandinavia, and sometimes even also Ireland. 'Peripheric' is thus commonly used with a spatial content and is strongly dependent on scale; it can be used at any scale, also at more national and regional levels. In its common understanding, 'peripheric' in the European context is also closely related to decision-making processes, the peripheric regions being those that are more distant and less influential in the processes leading to decisions that will also affect them, as happens in policy-making. In relation to peripheric, 'marginal' has a close, but slightly different meaning: 'marginal' also refers to a location, but mainly to uses and functions that are less important and dynamic in relation to the centre.

Abandonment

As to 'abandonment', the term is more closely related to marginalisation, and often the distinction between the two is not obvious. Nevertheless, a useful clarification may be to understand abandonment as a state, which may be the final outcome of a process, while marginalisation relates to the process itself. In this way, it can be said that a piece of land, or the agricultural use of the same, is abandoned or it is not. If there is still some kind of use, even if extremely extensive, the land cannot be classified as abandoned. And if the use changes from agricultural use to another, such as forestry, there may be abandonment of the agricultural use, but not abandonment of the land. The identification of abandonment of the land is complex: in some cases abandonment might be the result of an active decision – to stop practising agriculture or to leave a rural area for example – but in many cases abandonment will be the outcome of a more passive and gradual development following the lack of investment or renewal. In these latter cases it will be difficult to judge the actual moment or state of abandonment. Needless to say, even the state of abandonment does not have to be a final or irreversible one and can be changed when the contextual conditions change. Even abandoned land normally still belongs to a legal owner, who might decide to give it a use in the future even if this is not for agriculture but for another purpose. Thus, the time frame of analysis is a crucial issue. Furthermore, in relation to marginalisation, abandonment usually relates more strictly to the farmland or to the farming activity (Hunziker, 1995: 400; Moreira et al., 2001: 558), while marginalisation has progressively evolved to be seen as a more comprehensive process with several dimensions, as described in the introduction of this chapter.

In relation to abandonment, it is important to differentiate between situations such as spontaneous abandonment and planned withdrawal (Baldock et al., 1996). Increasingly in Europe, farmland is withdrawn from agriculture under schemes such as set-aside. Apparently abandoned land often is not truly abandoned, but merely temporarily out of use and awaiting a new owner or tenant. Besides that, particularly in marginal regions such as parts of the Iberian plains, due to natural constraints, arable land may be left fallow for many years, with only intermittent grazing. Or in silvo-pastoral areas, the use of the undergrowth may be abandoned, but with the oaks still being exploited for cork extraction, for instance. Over very large areas of Iberia, farmland is largely unfenced. Consequently, even if the land has been abandoned by its owner, it may continue to be grazed by sheep and goats from other livestock producers, thus preventing natural succession to woodland. Case studies in Portugal showed that the land may be covered by shrub and without use for agricultural production, but still not be seen as abandoned by its owner, as its cover is useful for other functions such as bee-keeping, game refuges, the prevention of erosion, and so on (Van Doorn and Pinto-Correia, 2006). The state of abandonment might thus not be as obvious or definitive as it seems (Baldock et al., 1996).

Furthermore, the issue of value has to be considered. Abandonment has a negative connotation since it is usually associated with social and economic decay (Derioz, 1991: 48; Liou, 1991: 57; Perez, 1990: 214). Therefore, it might sometimes be difficult to detect; in the Portuguese context, when dealing with statistical figures, there may be a substantial level of so-called 'hidden abandonment'. Alves et al. (2003: 10) explain how data from the General Agricultural Census in Portugal might conceal the reality of abandonment. Because of the lack of prestige associated with the abandonment of agricultural land, many farmers that have been questioned might still classify fields that are actually abandoned as agricultural land. In the Census data these areas might for example show up as 'poor pasture'.

Depopulation

Concerning rural depopulation, the term refers in the first place to a reduction in population density in rural regions, but it may have a broader dimension. The depopulation of agricultural and more remote rural regions has been registered in North-Western Europe and also in North America since the first decades of the twentieth century, when the terminology connected with marginalisation was not yet used. The most peripheric regions of Europe registered the same rural exodus some decades later, and the process has there been associated with marginalisation, of land or of the region. The process is still going on, even though in some regions this is accompanied by a growing outward movement from the cities to the rural areas. After decades of an almost constant increase of population in the countryside, depopulation resulted from two different factors: less need of labour in agriculture, and lack of economic strength and of alternative activities in the rural areas. As a consequence, the rural population started looking for new opportunities in other areas. Several studies emphasise the significance of a wider 'socio-economic health', where decrease and ageing of the population are followed by a deepening in the decline of the rural infrastructure, resulting in an accelerated downward spiral in local opportunities and facilities (Ilbery, 1998: 134). Depopulation then leads to a decrease in population, but often also to a reduction in the capacity of resistance to the forces leading to social, economic and also cultural decay (Bazin and Roux, 1995: 340).

Desertification

As to desertification, one can make a distinction between human and physical desertification. Although less used, the former generally refers to the depopulation and social draining of rural areas. The concept of desertification as a physical phenomenon is more common, generally defined as 'land degradation in arid, semi-arid and dry sub-humid areas resulting from various factors, including climatic variations and human activities' (United Nations, 1994: 5). Physical desertification is often related to the issues of drought, hunger and poverty in those regions where agricultural production is still essential to provide food and fibre for human survival, as in Africa and certain parts of Asia and of South America. As a consequence, research on desertification has also been focused mainly on soil and climatic issues as well as vegetation, land use and water management in these areas.

Recently, the scope of the scientific community has widened: not only by taking into account the processes in countries from the north Mediterranean and Eastern Europe but also through a regained focus on social dynamics in desertification processes. Leeuw (1999: 18) distinguishes two types of roles that people can play within a desertification process:

- 'actors', who actively contribute to the process through inadequate land use or bad agricultural practices; and
- 'observers', who react to decreases in soil fertility and reduction in agricultural yield by migrating from these areas.

Thus, physical and social approaches to desertification go hand in hand: reduced fertility and yield may encourage people to leave, and increased human pressure on fragile systems might lead to severe land degradation processes (Correia et al., 2004: 51). In Mediterranean Europe this cycle goes far beyond agriculture itself, including broader rural issues.

Marginalisation

As to marginalisation, several attempts have been made to clarify the concept, and a synthesis of the uses of the term 'marginalisation' in the literature can lead to its definition as 'a process, driven by a combination of social, political and environmental factors, by which in certain areas farming ceases to be viable under an existing land use and socio-economic structure' (Baldock et al., 1996; Chapter 1 of this volume). This relates to the farm level, and economic considerations by the landowner or farmer. However, a more detailed analysis shows that there may be various processes that can be connected with marginalisation, and that in the periphery of Europe there are more meanings to the term than in the north-western part of Europe.

Scale is an important issue: in the regions of more intensive and specialised agriculture, marginalisation usually refers to the reduction of economic activity in certain areas, locally or at the farm level, while farming may continue in the remaining area where conditions are more favourable. Locally, one can speak of extensification and marginalisation, but the general trend may be intensification and maintenance of competitive levels of performance; thus, the social impact of this marginalisation may be of little significance.

In peripheral regions such as the Mediterranean, marginalisation and a reduction or disappearance of intensity in agricultural use often affects whole regions, as has typically been the case in many of those regions classified as Less Favoured Areas (LFA) in the European context. In the affected regions, there may locally be poles of intensification, due to particular favourable conditions, but the dominating trend is extensification and thus marginalisation in the global context of European agriculture. These trends are often connected to social and economic decay at the regional scale, due to a decrease in economic activity and related depopulation, but this is not always the case. Sometimes other uses and new activities can replace agriculture.

Besides the issue of scale, there are several dimensions of marginalisation that need to be considered as they will also often lead to different consequences. To understand fully the phenomena that in scientific literature and in the political discourse have so far been connected to marginalisation, it is not only the economic but also the social dimension that has to be considered, and it is not only the maintenance (or not) of the agricultural activity in the land that has to be evaluated, but also its replacement (or not) by other uses.

In order to contribute to a more detailed understanding of marginalisation, the authors thus suggest the distinction of three processes. Each of them may be connected with one of the other concepts discussed above. The first two correspond clearly to two aspects of the concept of marginalisation presented above, as formulated in the first chapter of this volume:

- Type A. The marginalisation of land, where trends of use indicate that there is no direct economic use for the land in the present or near future, and that is often associated with (land) abandonment.
- Type B. The marginalisation of agriculture, where the activity ceases to be viable and becomes progressively less important or marginal in relation to other economic activities and also other uses of the land, but the land itself continues under a certain form of management and use.
- Type C. The marginalisation of the rural community, associated with depopulation, ageing, lack of dynamism and of innovation capacity, also recently often considered as a process of desertification and associated with peripheral situations (at various scales). This type is more what can be designated as social marginalisation, and does not make sense at field or farm level, but only at the community level.

The distinction made does not mean that these processes are always separated. On the contrary, they are often connected with each other: in the north-western regions of Europe, where intensive agriculture dominates, Types A and B may often coincide. Farming may stop in a former agricultural field, that remains without further intervention or management; this parcel may then be valued for functions such as nature conservation, landscape diversity or interest for recreation but it does not have an active management nor direct economic use. It will thus be subject to a process of both Types A and B. If the same field is planted with forest, or if because of progressive shrub development it can be more intensively used for hunting and managed with that purpose, for example, then it corresponds to a process of Type B only. The distinction is not always straightforward, but can be assessed through knowledge about the management and use of the concerned area.

In the periphery of Europe, the most common understanding of marginalisation as a mixed decay (a combination of processes), of the intensity in land use and of the dynamics of rural communities, does correspond to trends occurring in many areas at the local or regional scale. But these processes do not always go hand in hand, as the Portuguese reality shows. Therefore, a separated evaluation may lead to a clearer picture and thus a better understanding of the complexity of the processes taking place.

METHODOLOGICAL APPROACH: HOW TO DISENTANGLE MARGINALISATION?

In line with the distinction of processes as described in the previous section, a method is presented here that was applied to a study undertaken by the authors for the Portuguese Ministry of Agriculture (Pinto-Correia et al., 2006). The aim of this study was defined at its start as an evaluation of present marginalisation trends and effects in the Portuguese rural areas. Considering that there were various processes going on, that did not always follow the same path, and that consequently there was a need for clarification about what was going on where, each of the previously described processes of marginalisation was studied and analysed separately. The whole country (continental Portugal, excluding Madeira and Azores) was studied, at the level of analysis of the municipality. Thus, the focus was on dominating characteristics and trends at the municipal level, and not on farm-level or very local details. For each process a specific source of information was used, and specific indicators were selected. Each separate type of information was analysed through an integrated approach, based on the territorial distribution and dynamics of change, avoiding strictly sectoral or thematic perspectives. Furthermore, the objective was to focus on the existing situation, assessing the present characteristics and recent changes, rather than future risks.

The ultimate aim was to evaluate how the three processes are combined in each municipality. Each type of combination corresponds to a specific type of territory in relation to the dimensions of marginalisation and its effects – this evaluation could then lead to an assessment of potentialities or limitations in each municipality or group of municipalities. As described above, the processes considered were:

- marginalisation of land (Type A);
- marginalisation of agriculture (Type B); and
- social marginalisation, or marginalisation of the rural community (Type C).

To assess the dynamics, two moments in time were considered, according to the availability of data: 1990 and 2000. It would have been convenient to deal with data reaching a more recent period, but there was no information available for the selected level of analysis and indicators. The consideration of two periods in time allows estimation of the recent trends of change for each indicator, adding a dynamic dimension to the characteristics at a fixed moment, the main objective being to identify which types of marginalisation are occurring, at present, in the Portuguese territory.

For each of the processes, a range of indicators was selected and built up, using data from the different sources. For the marginalisation of land, the source used was CORINE Land Cover, offering data on land cover and its changes; two editions were used, from 1990 and 2000. For the marginalisation of agriculture, data from the Agricultural Census from 1989 and 1999, by the National Statistical Institute, were used. These data reflect the socio-economic dynamics of the agricultural sector as well as specific uses of land within the farm units. For the process of community marginalisation, the information used was derived from the Population Census, concerning 1991 and 2001, and also from the National Statistical Institute. The latter ones reflect socio-economic dynamics in a wider rural setting. The lists of indicators for the three dimensions are presented in Tables 2.1, 2.2 and 2.3. It is important to note that the analysis based on the selected indicators leads to the identification of various trends that may be positive, negative or stagnant, concerning the above-mentioned three dimensions of marginalisation. From the analysis, marginalisation trends can be identified, but positive dynamics or potentialities may also be detected.

The indicators in Tables 2.1, 2.2 and 2.3 correspond to different indexes built up on the basis of the available data: weight, variation and net change, and also persistence and swap:

- The weight is a relative index, which reflects the actual importance of the value for the indicator in relation to the value for a frame of reference, in a single time moment: for example, the share of agricultural population to the total population. This weighing of the indicators enables comparison between different municipalities in a single time moment and helps clarify the broad characteristics of each municipality.
- The net change, a concept coming from the field of spatial analysis, is also a relative index that reflects the difference in weight of an indicator between one year and another. It shows the relative dynamics between the two time moments, being the change evaluated in relation to a frame of reference that has to be the same in each moment of analysis.
- The variation, finally, is a non-weighed indicator, where changes are measured in absolute terms and the value from the second moment is compared to the value of the first moment; variation is useful as it helps to reflect major tendencies.

Table 2.1   Land cover characteristics and its dynamics in Portugal between 1990 and 2000 for seven clusters

	Dense urban areas	Peri-urban areas, agriculture under pressure	Dominantly forest	Degraded forest, stable agriculture	Stable agriculture and forest	Mainly agriculture, persistent	Contrasting dynamics
Persistence of the total land cover	85.0	84.8	83.0	91.2	78.2	91.8	89.1
Weight of aggregated urban land cover in 2000	64.5	2.8	2.7	4.4	24.6	2.8	1.4
Net change urban land cover	9.6	.8	1.0	1.2	8.6	.9	.4
Weight of aggregated agricultural land cover in 2000	19.2	21.9	29.8	47.9	40.7	78.0	47.8
Net change aggregated agricultural land cover	-8.1	-.6	-.2	-.7	-5.9	-1.0	-.5
Weight of aggregated forest land cover in 2000	7.0	56.4	28.3	34.9	18.5	10.6	10.7
Net change aggregated forest land cover	-1.5	.7	-3.5	-1.1	-3.0	.2	-1.1
Weight of shrub in land cover 2000	2.3	3.4	7.1	2.6	2.7	2.8	13.8
Net change of the shrub in land cover	-.3	-.1	-.8	-.1	-.1	-.2	-1.1
Weight of degraded forest, cuts and new plantations in land cover 2000	2.5	13.7	21.4	6.1	11.1	3.5	16.0

Table 2.1   (Contd.)

	Dense urban areas	Peri-urban areas, agriculture under pressure	Dominantly forest	Degraded forest, stable agriculture	Stable agriculture and forest	Mainly agriculture, persistent	Contrasting dynamics
Net change of degraded forest, cuts and new plantations in land cover	-.4	.3	5.3	.9	1.5	.4	2.2
Net change of irrigated cultures	-.2	.1	.1	.9	.2	1.8	.0
Net change of vineyards	.0	.0	.4	.2	.9	.9	.7
Net change of pastures	.0	.0	.0	-.1	.0	-.5	.1
Net change of agriculture with natural areas	-4.4	-.1	-.2	-.2	-2.5	-.3	-.7
Net change of natural pastures	.7	.0	-.2	.0	-.1	-.1	-.2
Net change of burned areas	.0	-.7	-1.1	-.1	-.8	.0	-.1
Net change of rice	.0	.0	.0	-.2	.0	.1	.0
Net change of orchards	.0	.0	.0	.0	-.2	.1	.3
Net change of olive yards	.0	-.1	.0	.0	.0	-.2	.0
Swap of shrub areas	.0	.0	.2	.0	.2	.1	.7
Swap of degraded forest, cuts and new plantations	.9	8.8	7.0	3.0	5.0	1.4	3.5
Weight of forestation between 1992 and 1998	.0	.5	.5	.9	.2	2.1	2.4

Table 2.2   Agricultural sector characteristics and its dynamics in Portugal between 1989 and 1999 for seven clusters

	Small farm units tending to intensification	Declining agriculture and progressing forest	Extensive agriculture in decay	Very large farm units with extensive farming	Diversity and dynamism in small farm units	Stability and important social role of farming	Medium and expanding farm units with extensive systems
Weight of agricultural population in 1999	14.4	24.8	30.7	14.6	12.1	56.8	23.5
Net change of agricultural population	-11.8	-16.1	-15.4	-3.5	-8.8	-7.4	-6.4
Weight of farmers in 1999	4.3	8.7	10.3	5.2	4.2	21.3	9.0
Net change of farmers	-2.6	-4.4	-3.7	-1.4	-2.8	-.5	-2.2
Weight of farmers over 55 years old in 1999	60.9	69.5	65.5	62.5	69.5	67.3	68.3
Net change of farmers over 55 years old	8.3	9.0	7.1	2.7	8.8	5.6	4.4
Weight of agricultural holdings with profits mainly from outside the holding in 1999	59.4	79.6	69.3	52.1	67.1	68.4	64.1
Net change of agricultural holdings with profits mainly from outside the holding	6.4	12.0	7.8	1.1	5.6	11.1	1.4
Net change of total surface of agricultural holdings in 1999	-14.6	-6.0	-7.6	5.1	-6.7	-2.4	6.4

Table 2.2 (Contd.)

	Small farm units tending to intensification	Declining agriculture and progressing forest	Extensive agriculture in decay	Very large farm units with extensive farming	Diversity and dynamism in small farm units	Stability and important social role of farming	Medium and expanding farm units with extensive systems
Weight of utilised agricultural area (UAA) within the farm in 1999	62.0	36.5	66.7	91.3	77.9	66.3	89.1
Net change of UAA within the farm	-.7	-7.8	-3.4	-3.3	-1.7	-5.8	.6
Weight of irrigable UAA in 1999	86.8	52.5	41.0	7.4	29.6	14.1	10.7
Net change of irrigable UAA	1.8	-6.1	-7.4	2.1	2.5	-2.6	.0
Weight of shrubs and forest within the farm in 1999	32.2	57.8	26.4	6.6	12.5	20.5	8.7
Net change of shrubs and forest within the farm	-1.7	7.1	1.9	3.0	-1.2	6.6	.4
Weight of poor pastures within the farm in 1999	2.0	3.9	19.7	25.3	4.8	6.8	32.0
Net change of poor pastures within the farm	1.3	2.4	9.3	16.6	2.8	1.3	23.4
Weight of non-UAA in 1999	2.4	3.9	3.8	.4	6.6	9.9	.9
Net change of non-UAA	.8	.2	-.2	-.2	1.5	-1.9	-1.0
Average size of the farms in 1999	4.9	7.5	7.2	127.1	9.9	10.4	53.2
Net change of the average size of farms	.7	1.7	1.4	32.6	2.8	.7	15.6

Table 2.3 *Community characteristics and its dynamics in Portugal between 1991 and 2001 for seven clusters*

	Urban and dynamic	Peri-urban dynamic	Young and industrial	Extremely strong decay	Strong decay	Stagnant and unqualified	Problematic but with some potential for resistance
Variation of the population	13.8	-17.0	-1.6	-7.2	5.0	14.7	-10.6
Weight of the population younger than 14 in 2001	16.1	8.8	14.7	12.9	18.8	15.8	10.9
Net change of the population younger than 14	-3.6	-3.2	-4.6	-4.4	-5.1	-3.9	-3.9
Weight of the population older than 65 in 2001	16.0	40.1	21.3	26.2	14.3	13.9	32.4
Net change of the population older than 65	2.3	7.4	3.6	4.8	2.4	2.9	6.0
Weight of the economically active population in 2001	48.8	32.8	42.2	39.9	45.5	52.3	36.2
Net change of the economically active population	4.8	4.2	2.8	2.0	2.7	4.0	2.8
Weight of the population employed in the primary sector in 2001	2.6	5.1	5.0	6.4	3.3	.7	6.1
Net change of the population employed in the primary sector	-2.9	-2.9	-4.7	-5.4	-3.7	-.5	-4.6
Weight of the population employed in the secondary sector in 2001	16.6	8.2	13.0	11.0	22.5	13.6	9.0

Table 2.3 *(Contd.)*

	Urban and dynamic	Peri-urban dynamic	Young and industrial	Extremely strong decay	Strong decay	Stagnant and unqualified	Problematic but with some potential for resistance
Net change of the population employed in the secondary sector	.2	1.2	1.7	1.7	.6	-1.8	1.6
Weight of the population employed in the tertiary sector in 2001	26.6	16.9	21.0	19.4	17.0	34.3	18.5
Net change of the population employed in the tertiary sector	7.0	5.6	5.5	5.3	4.1	4.3	5.0
Weight of the unemployed population in 2001	6.2	8.1	7.6	7.9	6.1	7.1	7.4
Net change of the unemployed population	.6	.0	.7	.5	1.3	.4	1.3
Weight of the illiterate population in 2001	9.2	26.8	13.9	18.0	10.2	5.9	20.6
Net change of the illiterate population	-2.8	-4.3	-3.1	-3.8	-2.0	-1.0	-4.3
Weight of the purchasing power PPC in 2001	89.7	47.0	63.8	55.7	59.0	130.1	52.4
Net change of the purchasing power	2.3	15.6	9.7	10.7	13.8	6.5	11.6
Ageing index (2002)	104.6	478.4	152.0	215.2	81.5	91.8	312.0

- Persistence, only used for the spatial data (Corine Land Cover) indicates the rate of area that was maintained under the same cover in the two moments of analysis.
- Swap, also only used for the spatial data, is the more complex type of index; while the net change shows the difference in area of one type of cover between one year and another, the swap reflects the area with that specific cover that has disappeared in one location but has been replaced in another location.

In a first phase, the indicators were analysed separately, through the calculation of the mentioned indexes. Comparisons between the various indicators enable some interesting interpretations and a detailed analysis of each municipality was done at this stage. A second phase was the cluster analysis, where selected indicators were combined and integrated in an automatic cluster calculation. The result corresponds to three separated clusterings, or typologies, of municipalities, according to each of the three dimensions considered. Then, the comparison between the three cluster analyses makes it possible to identify municipalities or groups of municipalities, where different combinations of processes are taking place in what concerns marginalisation.

DYNAMICS OF LAND COVER, AGRICULTURE AND RURAL COMMUNITIES: THE CASE OF PORTUGAL

In this section the results of the cluster analysis of the three above-mentioned dimensions are presented. Furthermore, some detailed analysis of municipalities with different characteristics in relation to these dimensions, and particularly to their combination, are described.

The first dimension concerns the analysis of the land cover. Table 2.1 presents the list of indicators selected, concerning both the present land cover pattern and the dynamics registered between 1990 and 2000. The table represents the corresponding average value per indicator, defined automatically through the cluster analysis for the group of municipalities in each cluster. Figure 2.1 shows the distribution of the municipalities within the seven clusters, in continental Portugal.

The analyses of the indicators on land cover, as well as the resulting map, show:

- A relatively strong dynamics of land cover change, affecting large parts of the country. The areas where agricultural land cover is more dominant tend to be stable, while the peri-urban areas are the most dynamic.

- A reduction of shrub areas and no signs of real marginalisation of agriculture in the rural areas. Marginalisation of agriculture can mostly be found in peri-urban areas, where most land will probably have a new use in the near future.
- Diverging trends, where locally some reduction of agricultural area is observed. However, there is also some increase in patches of intensive use (irrigated, vineyards), thus intensification, and an expansion of poor pastures, thus extensification, in rather large areas.
- Maintenance or increase of areas classified as degraded forest and shrub, and newly burned areas, probably indicating that the forest areas are those that are being abandoned – this affects mainly a large area in the centre of the country.
- Shrub and dense shrub, occupying large areas in parts of the country (for example in the north-eastern part of the country). They mainly cover areas with extreme physical conditions, where a productive use would be very hard to achieve, thus explaining the fact that they are very persistent.

The second dimension concerns the dynamics of the agricultural sector. The list of indicators together with the average values within the clusters is shown in Table 2.2. The distribution of the municipalities in the seven clusters is represented in Figure 2.2. Also here, the analysis concerns the characteristics of agriculture in each municipality in 1999 and the main dynamics between 1989 and 1999. The analysis of structural changes in the farming sector at national level shows:

- Some reduction of agricultural area, mainly concentration of farm units (due to a strong decrease in the number of farm operations), extensification of farming systems and local intensification.
- Signs of diversification of activities in the farm unit in some areas (north-east, north-west from Lisbon), but mainly of diversification in income, including income from outside the farm unit; this may reflect a survival strategy by farmers, or locally a vision of the coming multifunctionality demands – in any case, it corresponds to more diversification of rural areas.
- The most dynamic farming sector can be found in the north littoral, around Porto, in the area north-west from Lisbon and the large valley of the Tejo. Diversification is very remarkable in the north-eastern part of the country. Less dynamic and even decaying farming practices can be found in the forest-dominated areas in the centre and mountain regions.
- Stagnation and lack of innovation and extensification of agriculture are characteristics of the Alentejo region in the southern part of the country, even if with small differences between municipalities.

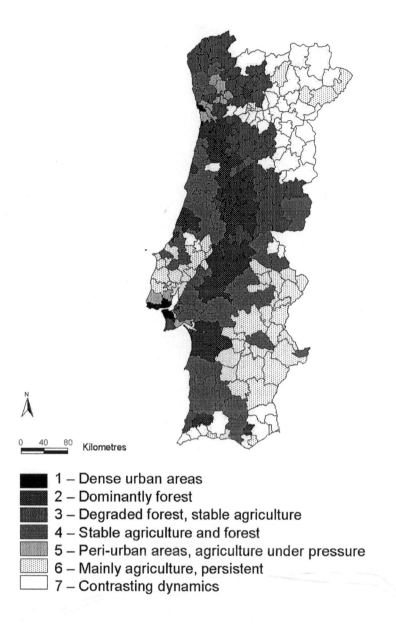

0   40   80   Kilometres

■ 1 – Dense urban areas
▨ 2 – Dominantly forest
▨ 3 – Degraded forest, stable agriculture
▨ 4 – Stable agriculture and forest
▨ 5 – Peri-urban areas, agriculture under pressure
▨ 6 – Mainly agriculture, persistent
☐ 7 – Contrasting dynamics

Figure 2.1   Distribution of land cover patterns in the seven clusters, in continental Portugal

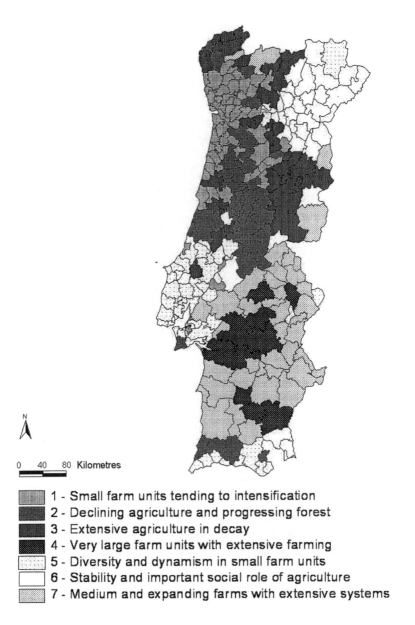

N

0   40   80 Kilometres

▓▓▓ 1 - Small farm units tending to intensification
███ 2 - Declining agriculture and progressing forest
███ 3 - Extensive agriculture in decay
███ 4 - Very large farm units with extensive farming
░░░ 5 - Diversity and dynamism in small farm units
☐ 6 - Stability and important social role of agriculture
▒▒▒ 7 - Medium and expanding farms with extensive systems

Figure 2.2   Dynamics of the agriculture sector in the seven clusters, in
*   continental Portugal*

Finally, the third dimension concerns the wider rural community, in what concerns mainly its social and economic characteristics. Here also a cluster analysis was developed, leading to seven clusters. The indicators included, as well as the final average values within each cluster, are presented in Table 2.3; the classification of the municipalities is represented in Figure 2.3. Some general conclusions can be drawn from the analysis of the socio-economic situation:

- The general trends of a dynamic coastal zone and a marginalised interior are maintained, and the contrast has even been accentuated. Some regional and sub-regional centres in the interior are poles of resistance and development.
- The dynamic and young areas around Lisbon and Porto are quite different from each other, the northern areas being younger, but less rich and more industrial, and the southern areas being more educated and more oriented towards services; they are nevertheless similar in size.
- In the Alentejo region there seems to be stagnation and very few opportunities, but also a reduction of unemployment. The decrease of unemployment rates in the Alentejo might be an indicator not so much of increased employment opportunities but rather of the fact that the unemployed have left the region.
- In the north-eastern part of the country, unemployment is high and the role of the primary sector still very important. Alternatives to farming are limited, but the region might have some potential for dynamism based on the multiple functions that the countryside has to offer.
- Some municipalities in the interior are found to be below what can be seen as a threshold of marginalisation, concerning all indicators, which makes it impossible under the present conditions to resist or change an advanced marginalisation process.

Different Types of Territories with Mixed Processes

Comparing the figures clearly shows there are various processes of change that nowadays occur across regions in Portugal. Often the different types of marginalisation coexist in one municipality, but often only one type can be found. Marginalisation of land according to Type A, where the land is in fact abandoned, remains limited. But there are many changes into other uses, like extensive farming systems, forestry and urban growth. In these latter two situations there is no marginalisation of land, but there is marginalisation of the agricultural activity, which has a progressively less important role, both economically and socially. This trend corresponds to the marginalisation process defined as Type B.

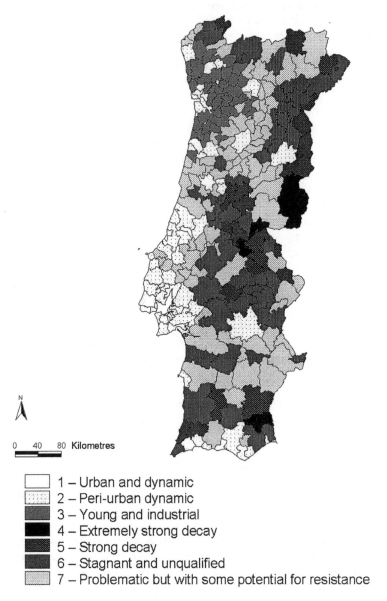

1 – Urban and dynamic
2 – Peri-urban dynamic
3 – Young and industrial
4 – Extremely strong decay
5 – Strong decay
6 – Stagnant and unqualified
7 – Problematic but with some potential for resistance

*Figure 2.3   Community characteristics and their dynamics in seven clusters,
in continental Portugal*

The data need to be interpreted carefully. If there is extensification of the farming system, it does not necessarily mean marginalisation of the land or of the activity – it can also mean a more adequate land use system in relation to the natural conditions of the land, and thus a more sustainable use in the long term. Furthermore, a general extensification at the regional or local scale is often followed by some trends of intensification at the local or farm scale, that is, in smaller areas. To evaluate this process of extensification and its consequences, an understanding of the context is important.

The clearest marginalisation process, the marginalisation of the land itself, of Type A, is occurring but not as a direct change after agricultural use. This type of marginalisation is probably occurring mostly in former agricultural areas which were planted with forest some decades ago, when agricultural use stopped being viable. Forestry was a possible use as it demanded a much more extensive care – but even this care has not been undertaken, as the owners have been too old or far away. This lack of management has resulted today in degraded forest areas, with high risks in terms of fire. Such forest areas can today be seen to be abandoned, with no management or care by landowners, and therefore in such areas one can talk of marginalisation of the land.

On the other side, social marginalisation, at a broad societal level, of Type C, affects large regions of the country, mainly the rural interior, where the capacity of resistance to decaying trends can be very low or non-existent. In some municipalities these trends are so strong that a possible shift towards a more active community is hard to imagine. Others show some potential for resistance, eventually in connection with the development of new functions in rural areas. But it is worthwhile noting that some municipalities which show a strong social decay, or marginalisation of Type C, may not show any signs of the other processes of marginalisation: no land being abandoned and/or no reduction of the agricultural sector. The farming systems tend to be extensive, and may be under further extensification, but this trend is explained and justified by the biophysical conditions, farm structure and policy context. And such farm systems based on extensive livestock breeding, in large estates, are also maintained when the landowner lives far away, and are based on very little labour. They can therefore be maintained with almost no relation or contribution to the local community – explaining how it is possible that there can be social marginalisation and no other type of marginalisation.

In another way, some municipalities showing decay in farming and an unstable land cover pattern – thus marginalisation of Type A and most commonly of Type B – reflect strong social dynamics connected with an expansion of the service sector and industries.

According to the presence or absence of the different processes mentioned, specific trends in various municipalities or groups of municipalities can be identified.

Alentejo interior
This includes municipalities such as Mertola, located close to the border to
Spain in South Alentejo, with a dry Mediterranean climate, very poor soils
and extensive land use and very low population density. The region largely
includes extensive farming based on livestock breeding, in large to very large
farm units, tending to extensification associated with environmental values
and potentialities for hunting and recreation. Some new patches of forestry
plantations emerged in the 1990s, but the landscape keeps its silvo-pastoral
character. There is hardly any marginalisation of land or of agricultural
activity; agriculture has maintained the traditional systems and is very
extensive in the European context. Still, it continues to exist and to be
generally well-adapted to the natural conditions and the farm structure.
Nevertheless, there is very low employment in farming, few alternative jobs,
and the already sparse rural community shows strong signs of decay – it is
clearly being marginalised socially and will hardly be able to resist social
decay.

Forestry areas in the centre
This includes municipalities such as Sertã, located in the geographic centre of
the country, with small plots, often dispersed in space. During last couple of
decades the region has faced a strong trend in the sense of agriculture
abandonment and forestry planting, with a large dominance of forest, that has
been replacing a small-scale and somehow archaic agriculture. Large areas of
forest are burned or degraded and show signs of very limited management;
the forest is marginalising and it is here that one can talk of marginalisation of
Type A, where the land is abandoned. The agricultural sector remains in
small-scale and traditional farms and is clearly also marginalised, showing a
low capacity for innovation. The community is not really dynamic, and two
possibilities can be proposed: further social decay and progressive
abandonment of villages, or maintenance of a small rural community living
from forestry combined with rural and environmental services.

Northern interior
This includes municipalities such as Mogadouro, in the north-eastern part of
the country, just north of the Douro valley, with diversified agriculture,
medium-sized farms, and medium population density, in relation to the
country average. The social indicators show low education levels and an
ageing population, but there are still active people and a strong relation to
agriculture. The farming sector is still diversified and specialised in regional
quality products, even if in small and medium-sized farm units and showing
trends of growing afforestation and extensification. These areas are broadly
marginal in relation to socio-economic development and also in relation to
agricultural economic capacity, but there is no significant marginalisation of

land or of the activity. Also, there are possibilities of a renewal based on rural services.

Central interior
This includes municipalities such as Sabugal, in the central part of the country close to the border with Spain, a clearly peripheral area according to all types of indicators, not least population dynamics. These are the most problematic rural areas of Portugal concerning marginalisation. Agriculture is limited, small-scale and old-fashioned. There is a dominance of forest that is not really managed, and the region has a very weak social community. All three dimensions of marginalisation are present in this region, and its future will certainly not be based on farming or on rural development and services.

Metropolitan areas of Lisbon and Porto
These are strongly urbanised areas, where the agricultural sector has been already marginalised – spatially, economically and socially. Farming occupies residual patches, progressively smaller and with less economic importance, the proportion of people working in the sector being extremely low. The role of production is negligible, but these few agricultural areas may have a role as green patches, being considered in the physical planning of these areas.

North-west of Lisbon
This includes municipalities such as Torres Vedras, 50 km north-west of Lisbon). In these areas there is no marginalisation of land, the agricultural sector or the community. There is low employment in agriculture, agricultural farming is maintained as a second activity, diversified and also profiting from the combination with rural services, and thus with high rates of income. The future pressure on land from urban areas might eventually lead to a marginalisation of agricultural activities.

Multifunctionality of Rural Areas or the Relativity of Marginalisation

According to previous reports and studies, large parts of the rural areas of Portugal can be characterised as less-favoured, fragile or marginal. Although many of these rural areas are indeed dealing with serious and complex problems, the characterisation of 'marginal' or 'fragile' is often based on a limited set of static indicators which offers little room for a more profound understanding of the dynamics and transformations that are really taking place in these areas. These same characterisations often do not take into account that different marginalisation processes do not necessarily coincide.

As marginalisation is a process rather than a state or an outcome, rural areas do not have a fixed position on a scale from favourable to marginal. Even the most marginal areas might in fact demonstrate signs of positive development, whereas less marginal areas might be marginalising. It is important to take these transformations and dynamics into account as they might be the basis for the future sustainable development of the area.

The same goes for the different perspectives on marginalisation: the function of food production (commodity outputs from farming) may be marginalised, but the same area may then support other functions considered valuable by society (non-commodity outputs), and thus the area will not be marginal from that point of view. Following the Organisation for Economic Co-operation and Development (OECD) terminology (OECD, 2001), multifunctionality corresponds to a combination of commodity and non-commodity outputs that can be produced in different degrees and combinations in different regions. Thus, an area that is marginal from an agricultural point of view does not necessarily have to be so from an ecological or recreation perspective, for example – it may even secure important functions of this other type. The opposite is also true, as the example from Portugal has shown. An area with little perspective for rural development, without an active and dynamic community, might still host a profitable type of agriculture. The analysis presented for the Portuguese reality has shown that for future land use strategies that really aim at securing and combining various functions in rural areas, it is important to disentangle these various levels of marginalisation.

Using the concepts of multifunctionality and territorial functions may offer an innovative analytical tool to evaluate these various processes going on, in particular in what are currently considered to be marginal areas of Europe. Much more than the concept of multifunctionality of agriculture, the concept of multifunctionality of the rural space makes it possible to consider agriculture through its wider role in the countryside, looking at the interrelations established between various dimensions, sectors and stakeholders (Cairol, 2005: 4; Wiggering et al., 2006: 240). This is a rather more encompassing definition of multifunctionality, available to be exploited by a much larger community of stakeholders. It is rooted in a reinterpretation of agriculture's contribution to rural development, its engagement with market processes and the changing role of farmers within a larger community of land managers (Potter, 2004: 19).

Multifunctionality of the territory can thus be defined as the capacity of the territory to support, or secure, several different functions that satisfy human needs, demands and objectives, in the broad sense. These can correspond both to commodity or non-commodity outputs of farming, in combination with the overall characteristics of the territory. The functions, that can be strictly connected or dependent on each other, can be classified as goods (removable) or services (not removable) (Brandt and Vejre, 2004: 10).

An analysis such as the one in the present chapter, where the characteristics and trends registered in the territories are assessed separately, may be the basis not just for understanding marginalisation in all its combinations, but also for evaluating limitations and potentialities for future functions to be offered in each type of territory. This assessment of multifunctionality corresponds to a theoretical and empirical challenge arising from the differentiated development of rural space in Europe (Marsden, 1998; Murdoch et al., 2003), both in its structure and in its uses (Ilbery, 1998).

CONCLUSIONS

This chapter has offered a new reflection on the concepts of marginalisation. The analysis of the processes occurring in rural Portugal has been considered as an adequate and rich example contributing to the understanding of the marginalisation concept and processes, since this country clearly belongs to what is seen as the periphery of Europe, where large areas are seen as marginal, both from an agricultural and also from a socio-economic perspective. Nevertheless, disentangling the concept of 'marginalisation', by looking at the various dimensions and combining static and dynamic analysis of indicators, shows that there are various specific connotations of this concept, and it therefore has to be used carefully. In other regions often considered as marginal in Europe, it may be that other dimensions are relevant, and those should be considered as well. What is most important in any case is to be aware of the complexity and diversity of what can be – by many authors and in many contexts – referred to and considered as 'marginalisation'.

Regarding the territorial approach to agriculture and the New Model of European Farming, there is a need for new insights to face new demands on rural areas and new roles of farming. The proposed distinction of processes corresponds largely to the different areas of attention that are currently being developed by the EU concerning the strategic guidelines and regulation for rural development by the European Agricultural Fund for Rural Development (EAFRD), including competitiveness of the agricultural sector, environmental quality and the social dimension. In this way, this distinction may also be a basis for defining priorities and strategies for differentiated investment and support for the future.

The typology that can be built from the separated analysis of the three processes may be a contribution to help focus on the specific drawbacks and potential of different territories, where agriculture and especially production agriculture is one function amongst others that may be as much or more valued. It should be clear that different territories might have a different vocation for certain functions. Focussing on the vocation of the territory implies the acceptance of a differentiation of the countryside.

Following the idea of differentiated rural areas does not mean that marginalisation in one dimension can always be balanced by valorisation of another dimension, or functions other than production. As has been shown above in the analysis of trends occurring in Portugal, in some areas a combination of various factors leads to a comprehensive marginalisation of the territory both from an economic and a social perspective. Still, even the comprehensive marginalisation processes can be seen as relative processes when taking into account other functions such as conservation, environmental quality, forestry production, or others – that are currently valued by society as well. Efforts of clarification of what is meant by marginalisation are thus extremely relevant in this debate, and they can only support more informed decision-making concerning the future of the rural areas concerned.

REFERENCES

Alves, A., N. Carvalho, S. Siveira, J. Marques, Z. Costa and A. Horta (2003), *Obandono da actividade agrícola*, Lisboa: Ministério da Agricultura.

Baldock, D., G. Beaufoy, F. Brouwer and F. Godeschalk (1996), *Farming at the Margins: Abandonment or Redeployment of Agricultural Land in Europe*, London and The Hague: Institute for European Environmental Policy (IEEP)/Agricultural Economics Research Institute (LEI-DLO).

Baptista, F.O. (1995), 'Agriculture, rural society and the land question in Portugal', *Sociologia Ruralis*, 35 (3–4), 309–21.

Bazin, G. and B. Roux (1995), 'Resistance to marginalisation in Mediterranean rural regions', *Sociologia Ruralis*, 35 (3–4), 335–47.

Belo Moreira, M. (2001), *Globalização e agricultura: zonas rurais desfavorecidas*, Oeiras: Celta Editora.

Benjamin, K., G. Domon and A. Bouchard (2005), 'Vegetation composition and succession of abandoned farmland: effects of ecological, historical and spatial factors', *Landscape Ecology*, 20, 627–47.

Bethe, F. and E. Bolsius (eds), (1995), *Marginalisation of Agricultural Land in Europe: Essay and Country Studies*, The Hague: National Spatial Planning Agency.

Black, R. (1992), *Crisis and Change in Rural Europe: Agricultural Development in the Portuguese Mountains*, Aldershot: Avebury.

Brandt, J. and J. Primdahl (eds) (1994), *Marginaljorder og landskabet marginaliserings debatten 10 år efter*, Forskningsserien 6, Copenhagen: Forskningscenter for Skov & Landskab.

Brandt, J. and H. Vejre (2004), 'Multifunctional landscapes: motives, concepts and perspectives', in Jesper Brandt and Henrik Vejre (eds), *Multifunctional Landscapes: Theory, Values and History*, Southampton and Boston, MA: WitPress, pp. 3–31.

Cabanel, J. and R. Ambroise (1990), 'La France part en friche ... et alors?' *Metropolis*, 87, 4–11.

Cairol, D. (ed.) (2005), *Multifunctionality of Agriculture and Rural Areas: From Trade Negotiations to Contributing to Sustainable Development. New Challenges for Research*, Summary of main results of the European

Project Multagri, Brussels: Commission of the European Commission, Sixth Framework Research Programme.

Correia F.N., C. Landeiro, I.L. Ramos, M.G. Saraiva, F. Nunes da Silva and F. Monteiro (2004), *Desertificação em Portugal, Incidência no Ordenamento do Território e no Ordenamento Urbano*, Vol. 2, Lisboa: Ministério das Cidades, Ordenamento do Território e Ambiente.

Derioz, P. (1991), 'Les conséquences spatiales de la déprise agricole en Haut-Langedoc occidental: l'éphémère victoire de la friche', *Revue de Géographie de Lyon*, 66, 47–54.

Gomez-Limon, J. and J. Lúcio (1999), 'Changes in use and landscape preferences on the agricultural-livestock landscapes of the central Iberian Peninsula (Madrid-Spain)', *Landscape and Urban Planning*, 44, 165–175.

Hunziker, M. (1995), 'The spontaneous reforestation in abandoned agricultural lands: perception and aesthetics assessment by locals and tourists', *Landscape and Urban Planning*, 31, 399–410.

Ilbery, B. (ed.) (1998), *The Geography of Rural Change*, Edinburgh: Longman.

Leeuw, S.E. (1999), 'Degradation and desertification: some lessons from the long-term perspective', in P. Balabanis, D. Peter, A. Ghazi and M. Tsegas (eds), *Desertification and Land Degradation in a European Context: Research Results and Policy Implications*, Vol. 1, Luxembourg: Publication of the European Commission, pp. 17–31.

Liou, V. (1991), 'Méthodes d'approches des friches dans le Parc Naturel Régional du Pilat', *Revue de Géographie de Lyon*, 66, 55–60.

Marsden, T. (1998), 'New rural territories: regulating the differentiated rural spaces', *Journal of Rural Studies*, 14 (1), 107–17.

Moreira, F., F. Rego and P. Ferreira (2001), 'Temporal (1958–1995) pattern of change in a cultural landscape of northwestern Portugal: implications for fire occurrence', *Landscape Ecology*, 16, 556–67.

Murdoch, P., P. Lowe, N. Ward and T. Marsden (2003), *The Differentiated Countryside*, London: Routledge.

Naveh, Z. (1991), 'Mediterranean Uplands as anthropogenic perturbation dependent systems and their dynamic conservation management', in O.A. Ravera (ed.), *Terrestrial and Aquatic Ecosystems, Perturbation and Recovery*, New York: Ellis Hoewood, pp. 544–56.

OECD (2001), *Multifunctionality: Towards an Analytical Framework*, Paris: Organisation for Economic Co-operation and Development.

Perez, M. (1990), 'Development of Mediterranean agriculture: an ecological approach', *Landscape and Urban Planning*, 18, 211–20.

Pinto-Correia, T. (1991), 'Land abandonment: changes in the land use patterns around the Mediterranean basin', *Cahiers Options Méditerranéennes, Vol. 1, Etat de l'Agriculture en Méditerranée*, Zaragoza: CIHEAM.

Pinto-Correia, T., B. Breman, V. Jorge and M. Dneboska (2006), *Estudo sobre o Abandono em Portugal Continental, Análise das dinâmicas de ocupação do solo, sector agrícola e comunidade rural, Tipologia de Áreas Rurais*, Évora: Universidade de Évora.

Pinto-Correia, T. and W. Vos (2004), 'Multifunctionality in Mediterranean landscapes: past and future', in R. Jongman (ed.), *The New Dimension of the European Landscapes*, FRONTIS Series, Wageningen: Springer, pp. 135–64.

Potter, C. (2004), 'Multifunctionality as an agricultural and rural policy concept', in Floor Brouwer (ed.), *Sustaining Agriculture and the Rural Environment; Governance, Policy and Multifunctionality*, Advances in Ecological Economics Series, Cheltenham, UK and Northampton, MA, USA: Edward Elgar, pp. 15–35.

Reenberg, A. and T. Pinto-Correia (1993), 'Rural Landscapes Marginalisation. Can general concepts, models and analytical scales be applied throughout Europe?' Leipzig-Halle, Germany: EUROMAB, 3, 3–18.

Romero-Calcerrada, R. and G. Perry (2004), 'The role of land abandonment in landscape dynamics in the SPA "Encinares del Rio Alberche y Cofio", Central Spain, 1984–1999', *Landscape and Urban Planning*, 66, 217–32.

United Nations (1994), *United Nations Convention to Combat Desertification*, New York: United Nations.

Van Doorn, A. and T. Pinto-Correia (2006), 'Refining the concept of land abandonment. Experiences from Southern Portugal', in Bob Bunce (ed.), *Mediterranean Landscape Ecology: Inside and Outside Approaches*, IALE Publication Series, pp. 147–58.

Wiggering, H., C. Dalchow, M. Glemnitz, K. Helming, K. Muller, A. Schultz, U. Stachow and P. Zander (2006), 'Indicators for multifunctional land use: linking socio-economic requirements with landscape potentials', *Ecological Indicators*, 6, 238–49.

3. Multifunctionality: what do we know of it?

Patrick Caron and Dominique Cairol

The concept of multifunctionality (MF) emerged at the end of the twentieth century and spread very rapidly because of non-sustainability concerns regarding the agricultural sector, marginalisation of agriculture or rural areas being one of these. The word has then followed different trajectories, some being conflictual, in its links with subsidies in the international trade negotiation arena (Blandford et al., 2003). The idea of MF emerged, indeed, as a co-construction between the political sphere and the research arena (Caron et al., 2008). It is difficult to say exactly when the term was first used. The word seems to have appeared in Austria in the 1980s, used by research mostly linked to environmental and landscape concerns (Brouwer, 2004). The interest in the word then grew as non-sustainability concerns were becoming more and more important, like food quality and farm diversification and pluriactivity. More specifically, research paid attention to the specific situation of Less Favoured Areas (LFAs), and studied the potential contribution of farm diversification as an alternative strategy for sustaining the livelihoods of (small) farmers and rural economies (Ilbery, 1988, 1991; Benjamin, 1994). Yet, the interest in ecological farming and low-intensity conventional agriculture and their impact on biodiversity and landscape diversity (Mander et al., 1999), and the necessity for marginal areas to promote different activities (Jervell and Dolly, 2003), address the links between MF and marginalisation. However, until the mid-1990s, references to MF remained quite confined.

MF basically refers to the fact that, besides the production of food and fibre, agriculture provides multiple services to society that are not necessarily compensated through the market system. This general definition rapidly leads to a polysemic concept, varying according to political contexts, academic disciplines and stakeholders. After a conflicting period linked to its use in the international trade negotiations, it has spread in terms of its meaning and use, and also geographically, since the late 1990s, based on dialectics between processes that related to the international trade negotiations on the one hand, and to the implementation of national and European political agendas on the other hand. The use of the word in research spheres has been strongly influenced by this rapid trajectory, mixing interest for what happened within society and the fear of dealing with a loose and fashionable catchword. This has probably prevented researchers from making it a stabilised concept.

A question remains whether MF can be used in the future to design strategies that aim to cope with marginalisation. To address this question, the first section of this chapter will come back to the circumstances of the emergence of the word. To that purpose, it will draw on the results of the Multagri project.[1] The objective of this project was to clarify the issues raised by the word of MF at the European level, through a state-of-the-art review of existing research and the identification of research gaps to construct a solid base for future research. This chapter reviews the different meanings of MF that can be identified. It will in particular try to clarify some ambiguity that relates to MF: is it a strategy or is it instrumental in analysing transformation? In addition, we will also discuss the usefulness of the concept to design strategies to cope with marginalisation.

EMERGENCE AND THE GROWING INTEREST IN MF[2]

The need to express how agriculture provides multiple services to society highlights the interest in promoting and sustaining these services. These are increasingly perceived and valued by society in high-income countries (OECD, 2002), but also in developing countries (FAO, 2002a, 2002b). Although the basic idea of promoting and supporting the multiple functions or roles of agriculture is not new – and has a tradition of several decades in Austria and Switzerland – the term 'MF' emerged in the late 1980s as an argument for continued government support to farmers, and spilled over to be used as a 'guideline' for agricultural policy reform in various countries, although varying from one country to another.

MF has been adopted in Article 14 of Agenda 21 by the plenary of the United Nations Conference on Environment and Development (UNCED) in Rio de Janeiro, Brazil, in June 1992, as a guiding principle for the Sustainable Agriculture and Rural Development (SARD) initiative. This principle explicitly calls for agricultural policy to be reviewed in light of the multifunctional aspects of agriculture, particularly with regard to food security and sustainable development. However, United Nations documents remain vague when it comes to the definition and operational content of both terms, MF and SARD. Some of them also avoid the term MF, because of the conflicting issues attached to it and relating to international trade negotiation in the late 1990s. This is the case with the Roles of Agricultural (ROA) project that was implemented by the FAO in developing countries in 2000.

The use of the word in international trade negotiations by the Friends of Multifunctionality as an argument to defend their subsidies actually coincided with the growing interest in research on this issue, whether the latter related to the trade negotiations or not. The political interest rapidly increased the research mobilisation, notably in view of the trade negotiations in the context of the World Trade Organization (WTO). By the end of the 1990s, the notion of MF had become quite controversial in the international arena.

By that time, the OECD had adopted the concept of MF as a policy principle, at the Ministerial Meeting of OECD Agriculture Ministers in March 1998, recognising that: 'beyond its primary function of supplying food and fibre, agricultural activity can also shape the landscape, provide environmental benefits such as land conservation, the sustainable management of renewable natural resources and the preservation of biodiversity, and contribute to the socio-economic viability of many rural areas' (OECD, 2001).

Likewise, in the Agenda 2000 programme for change in the Common Agricultural Policy (CAP) the European Union emphasises that:

> the content of the reform will secure a multifunctional, sustainable and competitive agriculture throughout Europe, including in regions facing particular difficulties. It will also be able to maintain the landscape and the countryside, make a key contribution to the vitality of rural communities and respond to consumer concerns and demands regarding food quality and safety, environmental protection and maintaining animal welfare standards. (European Commission, 2000)

This can be seen as an expression of changing societal concerns and a social and political answer to the problems associated with the monofunctional productivist agricultural model that had been promoted during the twentieth century. Hence, the objective of MF was officially featured in the CAP reform, and has been adapted by each EU member country into its own legislation. However, the interpretation of MF varies according to the specific circumstances in the different countries (Caron et al., 2008). It has been interpreted as a way of supporting and diversifying farming activities (for example in France and Italy), as a way to promote biodiversity and landscape management (for example in Germany and the Netherlands), as a way to favour rural employment (in Eastern European countries), or countries have remained reluctant to use the word apart from a restricted use for marginalised areas (for example England, Spain).

Following these policy initiatives, there was an increasing demand for academic expertise. First, recognising that the concept of MF is not well defined and is prone to different interpretations, the OECD mandated national reviews of literature and empirical studies in order to clarify its meaning. Academic work aimed to elaborate a common terminology, to identify key policy issues and to develop a framework for analysis[3] (OECD, 2001). Together with the demand for research on individual countries,[4] this induced further academic work on agriculture's MF and contributed to the spreading of the word into the academic sphere. However, this did not help to clarify the concept of MF. Rather, in response to political initiatives calling for expertise, various definitions and conceptions of MF have been proposed within the scientific community.

The term MF has firstly and predominantly been conceptualised by (agricultural) economists, as the OECD mandated research to elucidate the theoretical foundations and policy implications of the concept through economic approaches. But other disciplines also got involved, as controversies and interests were growing. As a consequence, the increasing use of the word led to debates and critiques concerning the relevance and applicability of MF as a policy concept, and forced the academic community engaged in MF research to position itself with respect to politics. Hence, it induced waves back from the academic world to the political arena and raised new research questions regarding the MF of agriculture and its relationship to sustainable development (SD).

THE DIFFERENT MEANINGS OF MF

As seen previously, the rapid spreading of the MF concept has made it both successful and highly controversial at the same time. Following its use in economics, researchers in sociology, agronomy, environmental sciences, geography, and so on started to adopt the notion and apply it to different subjects, such as farming systems, rural development and various societal movements. They saw in MF a continuity of programmes developed until then. Some of this research was triggered by the EU adoption of MF as a guiding principle of the CAP in Agenda 2000 and the resulting political orientations. However, much research was quite critical of the notion (Hediger, 2000), and researchers pointed out that the term MF had been put into political discussions without clarifying the concrete meaning and without relying on concrete information about adequate policy measures.

In order to deal with the wide diversity of conceptions of MF between countries and disciplines, a clustering has been devised by Caron et al. (2008) through a comparative analysis of national European case studies that aimed at grouping research practices. This classification does not intend to provide new definitions but to reflect upon the observed diversity, through the identification of Concept Oriented Research Clusters (CORCs). The latter are characterised by a relative homogeneity in the research practices, in the research questions addressed, in the concepts used or discussed by scientists to lead their work, and in the scientific disciplines, the stream of thought or possibly the epistemic community researchers belong to. This open classification is not a way of classifying countries since the national specificity of agricultural sectors does not fit with the way scientific communities are organised. Rather, the differences are due to different epistemological foundations and uses of the word MF or the use of synonyms instead.

The CORCs classification implies questions like the role and place of the agricultural sector in society, territory and regional environment, justification of public policies, integration of scales, territorial approach, and so on, in relation to the scientific origins of different approaches to address these questions. The clustering confirms that the focus does not always show the same level of abstraction in research. It encompasses theoretical or modelling work, empirical studies of MF practices and policy-orientated research. In addition, the level of maturity and stability is variable from one CORC to another. As in the trade-related context, the different authors share the same theoretical background, the same language and concepts, and generally stick to the political debate. The implications in terms of public policies are very often part of the scientific discussion. The distinction of eight CORCs is suggested.

CORC 1: A Joint Production of Commodities and Public Goods

This first CORC is built upon analyses of MF by neoclassical economists that started around 2000 in relation to the international debates on trade and domestic support to farmers. Researchers contributing to this CORC have adopted a shared and explicit definition of MF based on its aspects of jointness between commodity outputs and public goods and its aspects of externalities (OECD, 2001; Bonnieux and Rainelli, 2000). They often refer to the concept of non-trade concerns as a synonym of MF. This conception of MF is consistent with the 'positive' definition laid down and used by the OECD (2001). This cluster is fairly international, including American research, and uses a limited number of shared hypotheses and concepts arising from neoclassical economics (environmental economics, economics of production and trade or other sub-areas of welfare economics, neo-institutional economics, and so on). The related literature mainly focuses on the efficiency and legitimacy of public policies or institutional arrangements in order to promote joint public goods and positive externalities (Vatn, 2002; Hediger and Lehmann, 2003; Le Goffe, 2003; Romstad, 2004) and their legitimacy in relation to international negotiations. In that sense, the first CORC also involves a normative dimension, even if the definition of MF itself is essentially positive. Analytical firmness is the main strength of this CORC, and a main weakness is the lack of empirical evidence of jointness. One can, in addition, question the relevance and limits of such an approach when applied to situations where an informal economy is important, as is often the case in developing countries.

CORC 2: Multiple Impacts and Contributions from Agriculture to Rural Areas

The second CORC gathers interdisciplinary works focusing on the impact analysis of agriculture in a particular area. This cluster's originality is not the conceptual qualification of MF. It rather attempts to build an empirical and comprehensive focus of the state of agriculture in an area and its contribution to change (Laurent, 1999; SOLAGRAL, 1999; Berriet-Solliec et al., 2000; Léger, 2001; Pingault, 2001; Revel et al., 2002). This CORC concerns the contributions of agriculture to a local community, a region or society as a whole (Pilleboue, 2002; Bonnal et al., 2003). Findings on those aspects of MF are brought by economists, sociologists and agronomists addressing research questions such as the assessment of the impact itself (on employment, landscape, income, and so on), or how to promote farming diversification in agricultural and non-agricultural activities (important issue in Eastern European countries for instance: Hadynska and Hadynski, 2004). The empirical relevance of this CORC is its main strength for decision-making, whereas the lack of conceptual unity and robustness is its main weakness for research purposes.

CORC 3: A Complementary and Conflicting Connection between Commodities and Identity Goods

The third CORC mainly includes economists working on an alternative view of MF in reaction to the common definition. They do not share the dominant opinion that non-trade concerns in the field of agricultural MF should be analysed as a result of market failures, which would find a solution either by creating new markets or by adequate ways of public goods production. Researchers in CORC 3 consider that the development of market exchange unavoidably involves the destruction of identity and reciprocity structures, while the non-market exchange dimension of agricultural production might restore them. Empirical works of this analytical stream are conducted in several parts of the world (Barthélémy, 2003) and show how these two complementary and conflicting dimensions of agriculture are co-existing: on the one hand, market exchange organisations and market price systems; on the other hand, identity and reciprocity organisations and framing of non-market price systems. Researchers draw the concrete lesson that there always will remain two different (market and non-market) organisation and price systems, and that the political task consists in managing and controlling conflicts between them, and not to hopelessly keep trying to reduce one dimension to the other (Barthélémy and Nieddu, 2003). The main strength of this CORC is its ability to account for economical values in farm production that CORC 1 does not account for, in particular the cultural dimension. Its main weakness lies in its lack of anchorage into the standard economic literature and visibility in the policy debates.

CORC 4: Farmers, Strategies and Practices: Technical Change and Livelihood Systems

CORC 4 includes research of agronomists and economists who work at the farm scale and perceive MF as a motor that drives agricultural practices. It comprehends two different focuses:

- the design or the promotion of 'good practices' according to ecological norms; and
- the understanding of practices and farmers' individual choices by taking into account MF (Léger, 2001; Josien et al., 2001).

This CORC, and more particularly the second focus, enlarges the analysis of farming choices and decision-making processes. For economists, the interest lies in the way non-market objectives can be reached through private actors reacting to private signals. This requires new methods to assess and improve the procedure for farmers' decision-making, taking into account a wide range of functions and trade-offs. The basic research questions in this context relate to:

- the interpretation of MF in terms of farmers' decisions and behaviours; and
- the extent to which the recognition of MF (in public policies or in local institutions) has led to a change in farmers' practices and strategies.

The main aim here is not to qualify a list of functions of agriculture, but to consider these functions and their combinations (environmental protection, landscape management, family welfare, and so on) as factors of change and of producers' technical choices (Roep and Oostindie, 2005). This main strength of this CORC is its potential effectiveness in understanding and promoting principles of MF at the farm level. Its main weakness is the lack of a common analytical foundation.

CORC 5: Multiple Use of Rural Space and Regional Planning

CORC 5 encompasses work on MF as a policy guide to integrate new objectives in farm policies as a complement to the main drive towards agricultural modernisation and productivity. The normative dimension in this CORC is obvious. The aim is explicitly to provide a scientific basis for objectives such as redirecting funds to less-favoured areas, reinforcing the diversification of economic activity, and promoting alternative values of agriculture like landscape protection (Reig, 2004; Roep and Oostindie, 2005). As in CORC 2, and for the same reasons (empirical relevance), the conceptual roots of MF are not at stake. Research methods can be rather heterogeneous and research teams are pluridisciplinary, including scientists and experts from urban and rural planning, landscape architecture and social geography. Research in CORC 5 integrates multiple functions of agriculture but also multiple uses of the territory. A typical research question in CORC 5 is about the best way to organize spatial planning by taking into account the impact that agriculture may have on the attractiveness and sustainability of rural and urban living areas. This CORC is particularly well represented in the Netherlands, where competition between land users is high, but also in Spain where the dichotomy in looking at the agricultural sector is strong: either a wealth provider through specialized intensive production of a specific commodity, or a rural development issue and concern in LFAs. Its main strength is its direct orientation toward an evolution of policy-making. Its weakness, as far as research is concerned, is a lack of conceptual robustness in the definition of MF.

CORC 6: Adjustment between Activity Systems and Societal Demands as a Way toward Sustainable Agriculture and Rural Development (SARD) regulation

CORC 6 involves the work of authors who seize the emergence of MF as an opportunity to build a holistic view of agriculture as a way toward SARD, and therefore as a way to re-embed agriculture within society (Reig, 2004). The appearance of MF in the debate on sustainable development helps to point out the specific contributions of agriculture to rural development. This includes analyses of its role in food supply chains, of the compatibility between sustainable development and farm competitiveness, of its importance for the maintenance of rural population in less-favoured areas, and so on. Scientists belong to very diverse disciplines, but share the common concern of sustainability that goes beyond the analysis of functions and their relationships. The strength of this CORC is its comprehensive ambition, making it possible to analyse agriculture globally in the long run. Its main weakness is a lack of analytical firmness in the characterisation of agriculture.

CORC 7: A Social Demand Towards Agriculture

CORC 7 includes research focusing explicitly on the demand side for MF. Although the demand side is largely present in all CORCs, it is generally included as a given matter of fact in the other ones. For researchers in this CORC, MF is primarily defined by the multiple expectations or requirements of society toward agriculture (Bonny, 1999). Fundamentally, these expectations are the very justification for agriculture to be oriented in a multifunctional way. These authors developed methods to identify (Léger, 2001; Auvergne et al., 2000) and quantify these social demands (for example in terms of the willingness of taxpayers to pay) and, eventually, the ways agriculture might be able to meet them. The methodological challenge in this CORC is very high given the lack of reliable and objective information which is available and given the controversies on existing methods. The main strength of this CORC is the value of the information sought for policy-makers. For economists, its main weakness is the contradiction between the wide range of information required to evaluate the full non-market value of agriculture, and the level of precision required for these empirical econometric studies.

CORC 8: Governance, Policy and MF

CORC 8 is made up of research referring to the functions of agriculture that are explicitly and objectively recognised in legal or official texts underpinning agricultural policies. Researchers study the existence of MF in such texts and the consistency of new official objectives (Bodiguel, 2003, 2004) regarding the promotion of MF with the policy measures or the institutional arrangements implemented, for example the CTE (Contrat Territorial d'Exploitation) in France (Couturier, 2002). Other research questions are, for example: To what extent does MF modify the principles and modalities of previous farm policies? To what extent does it constitute a new paradigm (Massot, 2003; Delorme, 2004) or a new guide for agricultural policies (socio-economists, researchers in political sciences, jurists)? The main strength of this CORC is its ability to help judge whether political claims are actually converted into real policy reforms and farming practices, and to help provide an impact assessment of such policies. Its main weakness is a lack of analysis of the economic rationale of the policy measures.

In conclusion, while political recognition of MF at the international level seemed more and more difficult to obtain in the trade negotiation context, the concept of MF as a research object gained ground in all disciplines and streams. The explicit debate on MF is now located at national and local levels, and tends to move from a trade-related problematic to other fields like rural and agricultural development models.

Yet, the societal stakes raised by the political debate are still crucial at the international level. As a consequence of the liberalisation of agriculture, the need for agreeing on a multilateral framework for the design of national policies, that takes into account non-trade purposes, obviously calls for a renewed interest for looking at MF. This was confirmed by the reports from the World Bank (2007) and from the International Assessment of Agricultural Knowledge, Science and Technology for Development (IAASTD, 2008).

A better integration and strengthening of scientific communities is probably required. It is too early to call the identified clusters epistemic communities. The evolution of each concept of multifunctionality inside each epistemic community and the way research teams have made it a research topic is not yet completely clarified (that is, why and how some groups of researchers have used this idea of MF at some stage of their scientific work). This could further be addressed through specific research or networking.

But this migration from the political agenda to the research teams was probably a necessary stage for the idea of MF to gain more credibility, more objectivity and more scientific content, in relation to sustainable development.

A single definition of MF cannot be derived from the classification provided here. Although there might be a great deal of overlap between CORCs, the theoretical base upon which scientific production relies might be in some cases incompatible. This makes it impossible to think about one single integrated CORC. However, the state-of-the-art of research outputs and the diversity of meanings that have been identified for the word MF drive us to a new perspective for analysing the transformation of agriculture and rural areas. There is certainly a challenge in integrating partial knowledge derived from differentiated production processes and communities within the perspective of designing new comprehensive analysis.

MF AS A CONCEPT TO ANALYSE AND ADDRESS MARGINALISATION?

Before looking at the possible application of MF to design strategies to cope with the marginalisation of rural areas, one has to question the capacity of the word to offer elements for analysis and policy-making. Research that refers to this term is indeed in a consolidation process to gain credibility regarding the political discussions. Many of the debates emerge from the undifferentiated use of different understandings of MF, sometimes considering MF as an objective for policies, and in other circumstances regarding it as a fact. We choose here to consider MF as a fact, the understanding of which makes it possible to design an analytical framework and relevant policies. According to this option, MF can be considered an activity or outcome-oriented notion that describes the characteristics of farm production or diverse functions of land, focusing on relationships. The first part of this section will briefly describe the main component and principles of the suggested analytical framework, and the second will look at the relevance of this framework to address the marginalisation process through the notion of territory.

MF as an Analytical Framework

Several considerations lead us to restrict the use of the word MF to an analytical perspective:

- the discredit on the legitimacy of policies that acknowledge or address MF has hampered research to look at MF as a framework; this has particularly been the case within the UN Food and Agriculture Organisation (FAO) which rapidly abandoned the word as it was too controversial;
- the varied application of MF in the policies of the different EU countries has been considered by researchers as looseness and has made it difficult to design and adapt a common framework; and
- the confusion created by the distinction between normative and positive definitions proposed by the OECD, and of the real difficulty of isolating both dimensions, research communities have mixed the analytical perspective and the political implications of MF reference.

Such a position makes it possible to look at MF as a promising analytical framework, from the perspective of better understanding the transformation of agriculture and rural areas. Three justifications have been identified by Cairol et al. (2005, 2008) and Caron et al. (2008) for using MF to design such a framework:

- the interrelations between functions;
- the links between agriculture and society; and
- the relation with sustainability.

Functions and their interrelations
Among the different approaches identified through CORCs, there are strong differences regarding the functions considered and the way of considering the interrelations linking them. Although many studies have chosen to follow the sustainability concept by distinguishing three groups of functions (that is, economic, ecological and social), no list of functions can be considered as absolute and the relevance of each of them is highly contextual. Acknowledging that a single activity may simultaneously fulfil several functions is trivial. However, if the interrelation between functions is seriously taken into account, this profoundly challenges the analysis. Links between functions were already partly considered through agricultural systems approaches. Yet, MF puts these interrelations at the heart of the analysis.

Re-embedding agriculture in society
MF opens the field to more integrated analyses, in relation to the evolution of wider societal objectives (Cudlínová et al., 2005). Multifunctional agriculture can be looked at as a consequence of the changing needs and demands of consumers and society at large with regard to agriculture and rural areas. Most studies dealing with consumer and societal demands related to

agriculture and rural areas address its components separately: demand for quality food production; for environmental, ecological and landscape values; and for social and cultural aspects. However, there are clear correlations between these three dimensions of demand. Therefore, some researchers put forward that this demand is likely to be multidimensional in its nature, rather than directed exclusively to one dimension of agriculture and rural areas (Sautier, 2004).

Parallel to the evolution of demand, many farmers have engaged in new activities, through new strategies such as diversification and pluriactivity. Three directions are distinguished:

- deepening activities (adding more value to products, with organic farming, high-quality products, on-farm processing, short supply chains);
- broadening activities (development of new activities, such as management of nature and landscape, agritourism); and
- regrounding activities (pluriactivity or cost reduction through alternative use and valorisation of internal farm resources) (Ploeg et al., 2002).

As a consequence, the analysis of MF presupposes the inclusion of a much broader spectrum of organisational forms than the simple dichotomy between professional and non-professional farms (Renting et al., 2005). As agriculture is placed within a more global perspective and re-embedded within society, MF makes it possible to renew the ways of taking into account the links between agriculture and society through approaches in terms of networks and institutional arrangements (Murdoch, 2000; Renting et al., 2005).

MF as a pillar of sustainable development?
The relation between MF and sustainability is generally considered to be implicit and is rarely mentioned explicitly by researchers, often leading to confusion between both terms. There is usually a notable lack of scientific attention to the specific interrelations between these two concepts. As we saw before, MF can be considered an activity- or outcome-oriented notion (OECD, 2001) that describes characteristics of farm production or outcomes from lands, focusing on relationships. It lacks a direct or immediate temporal dimension. It can have a normative meaning, or be referred to within an analytical perspective. On the other hand, sustainability is a normative approach that has to do with society's wish and ability to preserve current consumption levels. It is a resource-oriented notion: it requires the maintenance of some aggregate measure of capital (stocks of physical or economic, natural, and social capital, and the possibility of trade-offs between them), in order to fulfil the needs of future generations. Thus, it has a clear temporal dimension.

Can MF, defined as such, help and bring some input to make development more sustainable? We consider that MF can provide a useful analytical framework that helps to make sustainability operational, in particular since it is based on activities and functions. This framework supposes clearly identifying and analysing the functions through activities, their combination and the social demand. The link between sustainability and MF can be made through the impact that activities may have on resources. Descriptors of the characteristics of the system should help to assess how the system can be modified and what could be the impact of changes. But the main contribution lies in the possibility it offers to look at a range of possible options and the way of addressing thresholds. In return, sustainability provides the criteria that are needed to make the analytical framework operational. Connecting MF to sustainability also requires taking into account the time and space dimensions. By understanding more about MF, it is possible to address sustainable development better.

MF Framework and Marginalisation

The use of the word MF often emerges because of marginalisation concerns. It then highlights more specifically the difficulty met by the agricultural sector in competing with other sectors, whether from an economic perspective or because of the preference of society for other land use patterns and the assignment of land to other sectors. In such situations, MF has sometimes been looked at as a strategy to revitalise rural areas that were under a marginalisation process. As in CORC 5 (multiple use of rural space and regional planning), the application of funds to less-favoured areas, the reinforcing of the diversification of economic activity and the promotion of alternative values of agriculture through economic and spatial planning were looked at as solutions to strengthen the contribution of agriculture to the attractiveness and the sustainability of rural and urban living areas. This has sometimes been the basis for discussing development models, their relevance and their applicability to different situations. This has, for example, been the case in the Netherlands (Roep and Oostindie, 2005), where CORC 5 reflects the opposition between the development of a segregation model (intensive and specialised agriculture in suitable areas) and the promotion of an integration model (multifunctional agriculture). In many Mediterranean areas, the abandonment of traditional subsistence systems was the result of uncompetitive farm structures, lack of alternative employment opportunities, ageing of the population, marginalisation and the collapse of traditional farming systems (Caraveli, 2000). However, the adoption of measures that aim at preserving low-intensity farming systems and at reducing marginalisation are sometimes looked at as sources of overexploitation of marginal land (for example overstocking) and further environmental degradation. This demonstrates the limits of a sectoral approach.

A territorial approach is sometimes looked at as a central approach to understanding and acting in an attempt to cope with marginalisation. For example it appears at the heart of the EU political agenda towards cohesion and against marginalisation. The integration of the territorial dimension into EU agricultural policies was achieved in a proposal of the European Commission (European Commission, 2003), with the objective of cohesion as the key challenge. At the Rotterdam informal ministerial conference in 2005, emphasis was made on achieving a consistent approach to the development of EU territories.[5] The territorial cohesion is supposed to establish a principle of equity amongst European citizens wherever they live. In particular, this means promoting the provision of equitable conditions with access to public services, and of optimal capacity for competitiveness to all its territories. This should particularly bear in mind the diversity of geographic or demographic conditions.

Apart from looking at general cohesion and equity among territories, the concept also provides a common spatial framework that makes it possible to look at the relationship between the sector and the social group where activities take place. The promotion of a better articulation between agricultural goods and services and new views and expectations by society is indeed usually required when agriculture is marginalised. This puts at stake the local dimension of agricultural activities and drives us to approaches which consider territory as the integration point between agriculture and wider society. Several CORCs dealing with MF look at this articulation, although each of them does so from a different and specific angle.

The territory also reflects quite well the transformation under process, particularly marginalisation, and therefore offers a relevant set of indicators. It is now well known that territorial identity is achieved through the different functions of productive activities (Becattini and Omodei Zoroni, 2003). Defined as a social entity, the territory finally relates to institutional arrangements, which represent the constitutive basis for change. These institutional arrangements provide opportunities for new activities and policies through renewed coordination process involving collective and public actions. This is clear for activities like direct marketing or agritourism, which evidently depend on the cooperation of consumers and tourists for the valorisation of products and services, but this is also the case for payments for public goods, or nature and landscape management, or farm-based care provisioning, whose successful articulation often critically depends on territorially based collective action, apart from the successful enrolment of extra-local state agencies.

As a consequence of the interest in territory as a way to look at marginalisation, and because of the links between territory and MF (Cairol et al., 2008), MF can be considered here as a useful framework. From an analytical point of view, work done within different CORCs can help in better understanding the relationship between agriculture and society and the territory transformation process by taking into account sustainability concerns as well as the consequence of change in terms of sustainability. It can also help in designing new activities and policies to reduce or prevent marginalisation, by promoting MF at the territorial level, as shown by the example of the French Territorial Management Contract (Contrat Territorial d'Exploitation – CTE) implemented through the 1999 Agricultural Act. The objectives of this new policy instrument were, among others, to increase the supply of high-quality products, and to protect natural resources, biodiversity and landscapes, through the design of contracts between government and farmers and the implementation of concrete actions. The ambition of the government was to sign 100 000 CTEs with farmers by June 2002, that is, one-sixth of farms. However, by November 2002, only 38 000 CTEs had been signed when the new government decided to replace the CTE by another contract, the *Contrat d'Agriculture Durable* or CAD (Sustainable Agriculture Contract), with a view to simplifying the procedures and reducing public spending.

CONCLUSIONS

Although it started receiving attention from the 1990s, the idea that agriculture has multiple functions is indeed nothing new. This had long been acknowledged. What is certainly recent is the need to give a name to this reality for policy-making reasons and perspectives. Consequently, discussion has gathered high momentum. The lack of a common definition and analytical framework and the intensity of controversies for ten years had put an international discredit on the political recognition of the concept of multifunctionality. Meanwhile, the use of the term has been promoted to address local political concerns that relate to marginal areas, within two different perspectives:

* to deliver payments for agricultural products and activities; or
* to stimulate the search for an alternative development model.

While one should stress the explicit political dimension of the word 'multifunctionality', its use by research and the mobilisation of various disciplines have provided new insights into rural transformations. Empirical evidence of what multifunctionality is about currently exists and is available through adequate documentation and publications. These research outputs are very useful at the very moment when the international discredit is reducing and when the notion of multifunctionality is once again gaining momentum (World Bank, 2007). Yet, a shortcoming is that it has remained largely case-study based. Some specific issues have also received little attention: limited information is provided on the socio-economic impacts of alternative and diversified farm activities in terms of their contribution to income and rural employment. Many issues are still to be explored in terms of synergies and spin-off effects as a result of the combined functions of a particular activity or the combined take-up of different activities.

The analytical framework for multifunctionality which has been sketched in this chapter brings new perspectives with which to assess agricultural activities and rural development processes in relation to their territorial context. The efforts made by scientists to characterize the combinations of functions that farming patterns provide or could provide in connection with societal demands are partial – but existing – answers to SD information needs. This should help in the future to articulate complementary knowledge and design strategies to cope with marginalisation.

However, apart from specific issues, or theoretical ones still to be explored, there still remains a lot to be known regarding multifunctionality. It is in particular necessary to assess policies that officially recognise multifunctionality as a central and structuring category. Such policies are recent and their impact has just begun to be analysed (Kröger and Knickel, 2005). This perspective should make it possible to address the debate that has been opened in the current chapter, that is the role that multifunctionality can play as a normative recommendation to achieve sustainability towards SD. Structuring knowledge and suggesting an analytical framework within a policy-oriented operational perspective represents, on the other hand, a key challenge. This could confirm the assumption made here that multifunctionality might bring added value to the study and mitigation of marginalisation.

NOTES

1. http://www.multagri.net.
2. This section draws upon the results of Work Package 1 of the Multagri Project, to present, compare, analyse and classify the different definitions and concepts used in political and theoretical discussions on MF. This team included: Denis Barthélemy, Patrick Caron, Ana Hadynska, Jakub Hadynski, Werner Hediger, Tristan Le Cotty, Henk Oostindie, Ernest Reig, Dirk Roep and Eric Sabourin.
3. The working definition of MF provided by the OECD (2001) includes two main elements: (1) the existence of multiple outputs being jointly produced, and (2) the fact that some of these outputs exhibit the characteristics of externalities or public goods, with the implication that markets are either lacking or operating deficiently.
4. The French government's interest in MF prompted, for example, the Ministry of Agriculture to commission some contributions both to non-governmental organisations (NGOs) (SOLAGRAL, 1999) and to economists from INRA (French National Institute for Agricultural Research) on the issue of the justifications of public intervention for the promotion of MF in view of the WTO agricultural trade negotiations.
5. 'The incorporation of the territorial dimension, as well as the concept of territorial cohesion add value to the implementation of the Lisbon and Gothenburg strategy

by promoting structured and sustainable economic growth (in EU Informal Ministerial Meeting on Territorial Cohesion (20/21.05.2005) in Luxembourg)'.

REFERENCES

Auvergne, S., B. Fallet and L. Rousseau (2000), 'Proposition d'une méthode d'aide à la concertation', Actes du Séminaire Cemagref–INRA.

Barthélémy, D. (2003), 'La multifonctionnalité agricole comme relation entre fonctions marchandes et non marchandes', *Les Cahiers de la Multifonctionnalité*, 2, 95–8.

Barthélémy, D. and M. Nieddu (2003), 'Multifonctionnalité agricole: biens non marchands ou biens identitaires ?', *Economie Rurale*, 273–4, 103–19.

Becattini G. and L. Omodei Zorini (2003), 'Identita locali rurali e globalizzazione', *La Questione Agraria*, 0 (1), 7–30.

Benjamin, C. (1994), 'The growing importance of diversification activities for French farm households', *Journal of Rural Studies*, 10, 331–42.

Berriet-Solliec, M., M. Guerin and D. Vollet (2000), 'Les défis de l'évaluation économique d'un dispositif à visée multifonctionnelle: le CTE', Actes du Séminaire Cemagref–INRA.

Blandford D., R.N. Boisvert and L. Fulponi (2003), 'Non trade concerns: reconciling domestic policy objectives with freer trade in agricultural products', *American Journal of Agricultural Economics*, 85 (3), 668–73.

Bodiguel, L. (2003), 'Le territoire, vecteur de la reconnaissance juridique de l'agriculture multifonctionnelle', *Economie Rurale*, 273–4, 61–75.

Bodiguel, Luc (2004), 'Multifonctionnalité de l'agriculture. Le droit rural à la confluence de la sphère marchande et des considérations sociales', CNRS UMR 6029, CRUARAP, Nantes.

Bonnal, P., M. Piraux, J.-L. Fusillier and D. Guilluy (2003), 'Approche de la multifonctionnalité de l'agriculture à la réunion: les modèles agricoles, la relation agriculture emploi et la perception des CTE par les acteurs locaux', Montpellier: CIRAD–TERA, Rapport final pour le MAAPAR.

Bonnieux, F. and P. Rainelli (2000), 'Aménités agricoles et tourisme rural', *Revue d'Economie Régionale et Urbaine*, 5, 803–20.

Bonny, S. (1999), 'Analyse des demandes adressées à l'agriculture', in H. Savy, O. Manchon and J. Racapé (eds), *Produire, Entretenir et Accueillir: La Multifonctionnalité de l'agriculture et le Contrat Territorial d'Exploitation*, Paris: GREP, pp. 55–65.

Brouwer F. (2004), *Sustaining Agriculture and the Rural Environment; Governance, Policy and Multifunctionality*, Advances in Ecological Economics Series, Cheltenham, UK and Northampton, MA, USA: Edward Elgar.

Cairol D., E. Coudel, D. Barthélémy, P. Caron, E. Cudlinova, P. Zander, H. Renting, J. Sumelius and K.H. Knickel (2005), 'Multifunctionality of agriculture and rural areas: from trade negotiations to contributing to sustainable development. New challenges for research', Multagri synthesis report (www.multagri.net).

Cairol, D., E. Coudel, K.H. Knickel and P. Caron (2008), 'Multifunctionality of agriculture and rural areas as reflected in policies: the importance and relevance of territorial view', *Journal of Environment Policy and Planning* (publication forthcoming).

Caraveli, H. (2000), 'A comparative analysis on intensification and extensification in Mediterranean agriculture: dilemmas for LFAs policy', *Journal of Rural Studies*, 16 (2), 231–42.

Caron P., E. Reig, D. Roep, W. Hediger, T. Le Cotty, D. Barthélemy, A. Hadynska, J. Hadynski, H. Oostindie and E. Sabourin (2008), 'Multifunctionality: refocusing a spreading, loose and fashionable concept for looking at sustainability?', *Internation Journal of Agricultural Resources, Governance and Environment* (publication forthcoming).

Couturier, I. (2002), 'La multifonctionnalité et le droit rural', Actes du Colloque SFER, *La multifonctionnalité de l'agriculture et sa reconnaissance par les politiques publiques*, Paris, 21–22 March.

Cudlínová, E., M. Lapka, J. Maxa, A. Dosch, J. Luttik, M. Miele, D. Pinduciiu, D. Sautier, M. Lošťák and H. Hudečková (2005), 'Consumer and societal demands for multifunctional agriculture: summary report comparing consumer and societal demand among different countries', Multagri WP2 Deliverable: D 2.1, http://www.multagri.net.

Delorme, H. (ed.) (2004), *La Polititique Agricole Commune: Anatomie d'une Transition,* Paris: Presses de sciences.

European Commission (2000), 'Agriculture's contribution to rural development', Discussion Paper presented at the International Conference on Non-Trade Concerns in Agriculture, Ullensvang, Norway, 2–4 July.

European Commission (2003), Council Regulation (EC) No. 1782/2003 of 29 September 2003.

FAO (2002a), 'Expert meeting proceedings. First expert meeting on the documentation and measurement of the roles of agriculture in developing countries', ROA Project Publication No. 2, Roles of Agriculture Project, Rome: Food and Agriculture Organization of the United Nations.

FAO (2002b), 'Socio-economic analysis and policy implications of the roles of agriculture in developing countries: Theoretical notes', Roles of Agriculture Project, ROA Team Leaders Meeting, 1–2 July, Rome: Food and Agriculture Organization of the United Nations.

Hadynska, A. and J. Hadynski (2004), 'Conceptions of multifunctionality: the state-of-the-art in Polish research work. National report for Poland', Multagri Research Program, Brussels, unpublished.

Hediger, W. (2000), 'Sustainable development and social welfare', *Ecological Economics*, 32 (3), 481–92.

Hediger, W. and B. Lehmann (2003), 'Multifunctional agriculture and the preservation of environmental benefits', Proceedings of the 25th International Conference of Agricultural Economics (IAAE), 16–22 August, Durban, South Africa, pp. 1127–35.

IAASTD (2008), *Executive Summary of the Synthesis Report of the International Assessment of Agricultural Knowledge, Science and Technology for Development,* Washington: World Bank.

Ilbery, B. (1988), 'Farm diversification and the restructuring of agriculture', *Outlook on Agriculture*, 17 (1), 35–9.

Ilbery, B. (1991), 'Farm diversification as an adjustment strategy on the urban fringe of the West Midlands', *Journal of Rural Studies*, 7, 207–18.

Jervell, A.M. and D.A. Jolly (2003), 'Beyond food: towards a multifunctional agriculture', Working paper 2003-19, Norwegian Agricultural Economics Research Institute, Oslo.

Josien, E., L. Dobremez and M.-C. Bidault (2001), 'Multifonctionnalité et diagnostics d'exploitation dans le cadre des CTE: approche méthodologique et enseignements tirés des démarches adoptées dans quelques départements', *Ingénieries*, 131–45.

Kröger, L. and K. Knickel (2005), 'Evaluation of policies with respect to multifunctionality of agriculture; observation tools and support for policy formulation and evaluation: summary report', Brussels: European Union, http://www.multagri.net/.

Laurent, C. (1999), 'Activité agricole, multifonctionnalité et pluriactivité', in H. Savy, O. Manchon and J. Racapé (eds), *Produire, Entretenir et Accueillir: la Multifonctionnalité de l'Agriculture et le Contrat Territorial d'Exploitation*, Paris: GREP, pp. 41–6.

Le Goffe, P. (2003), 'Multifonctionnalité des prairies: comment articuler marché et politiques publiques?' *INRA Productions Animales*, 16, 175–82.

Léger, F. (2001), 'Mise en œuvre territoriale de la multifonctionnalité de l'agriculture dans un échantillon de projets collectifs CTE', *Ingénieries*, 11–20.

Mander U., M. Mikk and M. Külvik (1999), 'Ecological and low intensity agriculture as contributors to landscape and biological diversity', *Landscape and Urban Planning*, 46 (1–3), 169–77.

Massot, A. (2003), 'Le paradigme multifonctionnel: outil et arme dans la renégociation de la PAC', *Economie Rurale*, 273–4, 30–44.

Murdoch, J. (2000), 'Networks: a new paradigm for rural development?' *Journal of Rural Studies*, 16, 407–19.

OECD (2001), *Multifunctionality: Towards an Analytical Framework*, Paris: Organisation for Economic Co-operation and Development.

OECD (2002), *Agricultural Policies in OECD Countries: Monitoring and Evaluation,* Paris: Organisation for Economic Co-operation and Development.

Pilleboue, J. (2002), 'Quand l'expression de la multifonctionnalité de l'agriculture passe par la construction d'une image territoriale: le cas de l'Aubrac', Actes du Colloque de la Société française d'Economie Rurale, *La multifonctionnalité de l'agriculture et sa reconnaissance par les politiques publiques*, Paris, 21–22 March.

Pingault, N. (2001), 'Une évaluation multicritère pour des politiques multifonctionnelles', *Notes et études économiques*, Ministère de l'Agriculture et de la Pêche, Paris, France, 14, 51–69.

Ploeg, J.D. van der., A. Long and J. Banks (2002), *Living Countrysides. Rural Development Processes in Europe: The State of the Art*, Doetinchem, the Netherlands: Elsevier.

Reig, E. (2004),'WP 1 Multagri national Spanish report', http://www.multagri.net/.

Renting, H., H. Oostindie, C. Laurent, G. Brunori, A. Rossi, M. Charollais, D. Barjolle, S. Prestegard, A. Jervell, L. Granberg and M. Heinonen (2005), 'Multifunctionality of activities, plurality of identities and new institutional arrangements', Multagri WP4 Deliverable: D 4.3 http://www.multagri.net.

Revel, A., B. Roux, P. Bonnafous, B.-T. Ly and E. Fiack (2002), 'Multifonctionnalité des systèmes diversifiés dans les exploitations agricoles du Languedoc-Roussillon', Actes du Colloque de la Société française d'Economie Rurale, *La multifonctionnalité de l'agriculture et sa reconnaissance par les politiques publiques*, Paris, 21–22 March.

Roep, D. and H. Oostindie (2005), 'Agricultural multifunctionality: the state-of-the-art in Dutch research work. Capitalisation of research results on the multifunctionality of agriculture and rural areas', Sixth Framework Research Programme, http://www.multagri.net.

Romstad, E. (2004), 'Policies for promoting public goods in agriculture', in F. Brouwer (ed.), *Sustaining Agriculture and the Rural Environment: Governance, Policy and Multifunctionality*, Cheltenham, UK and Northampton, MA, USA: Edward Elgar, pp. 56–77.

Sautier, D. (2004), 'Consumer and societal demand of multifunctional agriculture in France', Multagri French Country Report. WP2.

SOLAGRAL (1999), 'La multifonctionnalité de l'agriculture dans les futures négociations de l'OMC', Rapport d'étude pour le Ministère de l'agriculture et de la pêche, Nogent sur Marne: SOLAGRAL.

Vatn, A. (2002), 'Multifunctional agriculture: some consequences for international trade regimes', *European Review of Agricultural Economics*, 29 (3), 309–27.

World Bank (2007), *World Development Report 2008: Agriculture for Development*, Washington: World Bank.

4. The clock is ticking for rural America[1]

Michael Duffy

The United States is a large and diverse country. US agriculture and rural America are equally diverse. As such, it is hard to make general statements about the country and what is happening. But, there are some observations that can be made regarding the changing face of rural America. The United States has become a largely an urban nation. Approximately three-quarters of the population lives on just 3 per cent of the land base. In spite of, or perhaps because of, the urban population concentration, there is a growing concern over the loss of farmland and rural open spaces. Social and economic changes are never neutral in their impact. These changes produce gains for some and losses for others. When evaluating changes it is important to consider more than the net impact of the changes. In other words, the impact of change is not just the total of the benefits and costs, but also the distribution of those benefits and costs. The impact of a change is also determined by who are the winners and losers.

All rural communities in the US are undergoing changes. The rural, non-farm communities in the US are undergoing changes in the population mix and the industrial base used for economic activity. In addition there are occupational shifts and changes in the lifestyle choices made by the residents of these non-farm communities. The changes in the agricultural communities are primarily associated with the changes occurring in production agriculture. In addition there are changes with the introduction of more agricultural processing industries closer to the sources of production.

The interface between the non-farm rural and the agricultural communities is where we are seeing considerable friction. The emergence of a dual production agriculture sector is a major contributor to this friction. Using Iowa as an example, the state's population mix has changed to such an extent that there are now more people living in the country but not on a farm than there are on farms. This change, coupled with the changes in production agriculture, has created a dynamic situation with considerable tension between the agricultural and the rural non-farm communities. This tension between the two communities arises for two fundamental reasons that are discussed in more detail later in the chapter.

The first reason for the tension is the encroachment of urban areas towards farming areas. This encroachment and the increasing non-farm rural populations means that as farmers go about their normal business there is more chance for interference with the non-farming population. Moving equipment, spraying land, and other farming activities stand more chance of interfering with other people the higher the non-farm population (Nickerson, 2001).

Another reason for the tension is the changing nature of agricultural production in the United States. The United States is moving towards a more large-scale production system. This is especially true with livestock. As the concentration of animals increases, so too do the problems with odour, potential environmental problems, flies, and other problems. There are many studies and papers on this aspect of the tension between farming and the non-farming community, especially as the non-farming community moves closer to the farms (Duffy, 1994; Swine Odor Task Force, 1995; Tyndall, 2006).

This chapter examines the issues of marginalisation as they relate to circumstances in the United States. Marginalisation as used in this chapter refers to situations where farming is no longer a viable activity given the existing patterns of land use and the changes in socio-economic structure. Before discussing the changes in rural America and the programmes that are being used to protect rural America, it is necessary to examine the changes that are occurring within US agriculture and that have an impact on rural America. The results of these changes are more pronounced in agricultural regions.

CURRENT SITUATION

Changes in Agriculture

US agriculture is characterised by increasing concentration in production, processing and retailing. In addition, there are increases in vertical integration of the production agriculture sector with more value-added processing occurring.

A farm in the United States is defined as any place that sold or could have sold $1000 in agricultural products annually. Based on the 2002 Census of Agriculture, there were approximately 2 million farms in the United States.

Figure 4.1 shows the distribution of farms, sales and land based on sales class. The majority of farms are very small in terms of sales. Notice that the very smallest farms, those farms that could have but did not sell $1000 worth of agricultural products, represented 27 per cent of all US farms, but had only 0.5 per cent of the sales. These farms controlled approximately 9 per cent of the land in farms in the US.

Figure 4.1 illustrates the bimodal nature of US farms. The very small farms represent the majority of farms but have a very small percentage of the sales. The largest farms, the 15 per cent of the farms with sales greater than $100 000, reported 88 per cent of the total US sales. The distribution of land is somewhat different than the distribution of sales. Figure 4.1 shows that land is distributed more equally among the farms. The top 15 per cent of the farms that reported 88 per cent of the sales had only 55 per cent of the land.

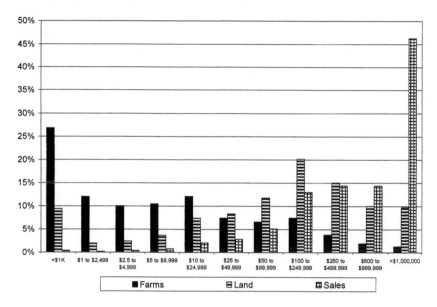

Source: USDA/NASS (2004).

Figure 4.1 Percentage of US farms, sales and land by sales group, 2002

The trend towards a dual agriculture in the US appears to be accelerating. Figure 4.2 presents the changes in farms, sales and land by sales class between 1997 and 2002. The smallest category farms, those who could have sold $1000 worth of agricultural products but did not, increased by 37 per cent, and the very largest farms, with sales over $1 million, increased by 8 per cent. The numbers in all other farm size categories decreased. The increase in the very largest and the very smallest farms represents what has been termed 'the disappearing middle' of US farms. In other words, we are seeing the middle-sized farms change in size or go out of business entirely.

The US Department of Agriculture (USDA)'s Economic Research Service (ERS) has devised a typology to help illustrate the nature of production agriculture in the United States (Hoppe et al., 2000). In this typology, farms are divided based on the amount of sales, the stated principal occupation of the operator and the financial characteristics of the farm.

The USDA typology breaks farms into eight categories. Five of the categories are for small, family farms. The USDA definition of a small farm is a farm with sales less than $250 000 annually. The small farms include limited resource farms (small farms with a limited asset base and income), retirement farms, residential or lifestyle farms, lower sales farms (farms with sales less than $100 000 but still farming is listed as the principal occupation), and higher sales farms (farms with sales between $100 000 and $250 000 and farming is the principal occupation). Two of the remaining categories are large family farms. The large family farms have sales between $250 000 and $500 000 and the very large farms have sales greater than $500 000. The remaining category is for non-family farms of any size. These farms are generally owned by corporations or cooperatives.

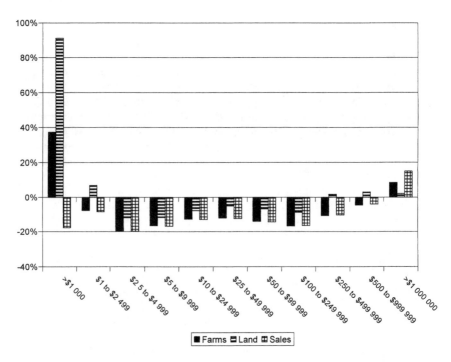

Source: USDA/NASS (2004).

Figure 4.2 Percentage change in US farms, sales and land by sales group, 1997–2002

The residential or lifestyle category farms represented 40 per cent of US farms in 2004. These are small farms (farms with sales less than $250 000) where the operator reported something other than farming as the principal occupation. Another category, retirement farms, represents 16 per cent of the farms. These are small farms where the operator reported being retired. These two categories account for over half the farms in the United States.

The Census numbers and the USDA typology show that a large majority of US farms are small farms. These farms control a relatively small percentage of the value of land and buildings in the United States. For example, the 56 per cent of the farms that are lifestyle or retirement farms control just 11 per cent of the total value of lands and buildings in US farms (Hoppe et al., 2000).

The structure of production agriculture has changed considerably in the US. Today, US agriculture can be classified as a system where the majority of the farms are small farms. These small farms produce a small fraction of the value of US agricultural production. The small farms do not, generally, produce enough income or provide full-time employment for a family.

The changing structure of production agriculture is important to remember when considering the marginalisation of agriculture in the US. This changing structure has a substantial impact on the efficacy of the programmes. If a programme is to reach the most people then it must be designed to appeal to the small farms. But, if a programme is intended to impact upon the most production then it will be necessary to attract the large farms. The situation is not so clear if a programme is intended to influence land and land use. The majority of land is held by farmers in between the two size groups. And, as shown in Figure 4.2, these are the farmers that are undergoing the most change. This makes designing programmes for them problematic.

Coupled with the dramatic change in the characteristics of production agriculture has been a dramatic change in the input supply industry. Seed, chemical and fertilizer manufacturers and retailers have undergone substantial consolidation over the past several years. These industries, as well as processors, are characterised by fewer firms and a much higher concentration ratio of the sales by the top four firms (Heffernan and Hendrickson, 2001; King, 2001; Fernandez-Cornejo, 2004).

Another recent phenomenon that influences farming is the tremendous increase in the value of farmland. The average value of an acre of US farmland was $1360 in 2004 (Barnard, 2006). This was over double the value in 1987 and 25 per cent higher than the value in 2000.

There are many reasons for the increase in land values. The relatively sluggish performance of the US economy has caused many investors to look for alternative forms of investment, such as land. The lower interest rates have also increased investor interest in land. The current US farm programmes continue to support land prices. And the final factor influencing land values has been the demand from non-farm uses: 'Recent research indicates that nonfarm influence accounts for 25 per cent of the market value of US farmland' (Barnard, 2006: 13). The non-farm uses include urban expansion and the demand for land for recreational purposes. Urban expansion continues in the US: 'urban area increased about 7.8 million acres (13 per cent) from 1990 to 2000' (Lubowski et al., 2006: 28).

Changes in Rural America

Against this backdrop of the changing structure of US agriculture, we also are seeing changes in the rural communities. The United States is divided into 3141 local government units called counties. Almost two-thirds, 65 per cent, of the counties are classified as non-metropolitan. There are 21 per cent of the counties in the United States that are classified as completely rural with populations of less than 2500. The metropolitan counties, 35 per cent of the total, have 83 per cent of the US population.

Only 14 per cent of all US counties, or 21 per cent of the non-metropolitan counties, are classified as farming dependent. A farming-dependent county is one where 15 per cent or more of the earnings in the county is from farming. The number of farming-dependent counties has been dropping significantly over the past several years. From 1989 to 2004 the number of counties designated as farming-dependent dropped by 28 per cent.

The population of rural America is also changing. The total US population grew by 13 per cent from 1990 to 2000. Not all counties showed an increase in population. The non-metropolitan and especially the farming-dependent counties showed a loss. This was especially true for the farming-dependent counties that rated low on a natural amenities scale (McGranahan and Beale, 2002).

Another significant change occurring in both agriculture and non-farming communities is the ageing population. Data for Iowa best illustrate this issue. Figure 4.3 shows the percentage of Iowa farmers under 35 and over 65. In the early 1990s Iowa shifted to having more farmers over 65 than under 35 years of age. At the last Census, in 2002, only 7 per cent of the farmers were under 35 and almost 25 per cent were over 65. The ageing population is also revealed by the age of the landowners. In Iowa almost half of the land is owned by people over the age of 65. As shown in Figure 4.4 the percentage of land owned by people over 75 doubled from 1982 to 2002 (Duffy et al., 2004). The increases in age are greater in the rural counties. In 1991, 16 per cent of the people in metropolitan counties in the US were over the age of 60.

That same year 19 per cent of the non-metropolitan, rural, county population was over 60. Just ten years later the percentage of people over 60 had dropped in the metropolitan counties while the share of people over 60 had increased in the non-metropolitan counties. By 2001 almost one in five persons in a rural county was over the age of 60.

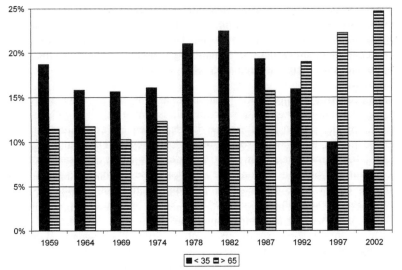

Source: United States Census of Agriculture.

Figure 4.3 Percentage of Iowa farmers over 65 and under 35 years old

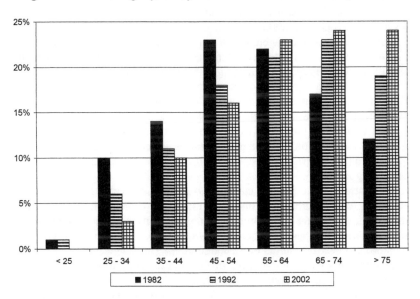

Source: Duffy et al. (2004).

Figure 4.4 Percentage of Iowa farmland by the age of owner

The overall situation in the United States reflects a decreasing number of farms. There is an increase in a dual agricultural production system and an increase in part-time or retirement farms. There is a considerable decrease in the number of counties dependent upon agriculture for their income. And, coupled with this, there has been a decrease in the economic viability of many rural counties.

The economic vitality of the rural community is critical when evaluating the marginalisation of agriculture. Communities that are ageing, with stagnant or declining populations, and are heavily dependent on agriculture, are less likely to attract new businesses or residents. They will be more likely to succumb to the marginalisation of agriculture. This is not necessarily true for all communities but it is certainly the case for those communities without the ability to attract new businesses. For example, in Iowa two-thirds of the counties lost population between 1990 and 2000. Several of the surrounding states are also experiencing infinitesimal to no population growth.

Marginalisation

So what do these changes imply for marginalisation of US agriculture? The USDA's Economic Research Service has done a number of studies examining conditions in rural America. It has produced numerous reports examining many aspects of these issues (Reeder and Brown, 2005; McGranahan and Sullivan, 2005; Stenberg, 2002).

One of the issues associated with the marginalisation of agriculture is the loss of farmland. In the United States, the severity of this problem depends to a large extent on the region of the country being examined. The entire United States experienced a 5 per cent decrease in the amount of land in farms from 1990 to 2004. Over the same time period, three areas experienced more than a 10 per cent decrease in the land in farms. These areas were the north-east, the south-east, and the Mississippi Delta states. Although these areas experienced a substantial decline in the land in farms, it is interesting to note these states only comprised 10 per cent of the total land in farms in the United States during 2004.

Urbanisation is the biggest cause of marginalisation of US agriculture. Land values for both urban land and farmland are at record high levels. As the communities grow there is pressure to convert land from agriculture to urban land uses. The major metropolitan areas within the US are the areas with the greatest impact of urbanisation on agriculture.

Land abandonment is not a major problem in the United States. However, it is important to keep in mind that the United States is a very diverse country and that some areas are facing different issues with respect to land use and land use policies. In addition, there also is a strong demand for recreational land. This land is used primarily for hunting, sightseeing or, in some cases, for country homes.

The changes in the nature of production agriculture contribute to one aspect of marginalisation of agriculture in the US. Changes in production can be characterised as substituting capital for labour. A result of this substitution has been increasingly tight margins for farmers. The tightening margins have led to larger farms sizes to generate income. The larger farm sizes have in turn led to loss of rural populations and a decline in economic activities in rural areas, especially those areas dependent on farming.

Figures 4.5 and 4.6 illustrate the nature of these changes. Figure 4.5 shows the increase in the value of the output for the US. The figure also shows the increase in the total expenses to produce the output. The difference is the net farm income, which is also shown in the figure. Figure 4.5 illustrates the changing nature of agriculture in the US. The value of the output has increased, reflecting for the most part the increases in productivity. But notice that the increase in yields comes with a price. Costs have been increasing at almost the exact same rate as output. Notice that the slope of the lines are almost identical. As the value of the output increases and as the cost of production increases, the farmer is left with essentially the same net income. The net income line in Figure 4.5 has a slight upward slope but not near the rate of the output or cost.

Figure 4.6 shows the net income as a percentage of the gross value of the output. This shows the steadily decreasing margins associated with production agriculture in the US. In the 1950s net income averaged approximately 36 per cent of the gross income. From 1995 to 2005 this average dropped to 21 per cent. This means that in the 1950s for every $100 of gross the farm kept approximately $36; today it is only $21. The farms have to be bigger to make the same income. In addition, as will be discussed shortly, the government payments have become a larger part of the gross output. Without these payments the margins would be even tighter. A natural result of the increasing farm size is a decreasing number of farms and farmers, especially in areas dominated by land-intensive commodities.

Recently in the US there have been movements away from the intensive, tight-margin production. Consumers have been seeking food and fibre attributes beyond the cost of production. These attributes, such as organic, non-GMO (genetically modified organism), natural, animal-friendly, locally grown, and so forth, are increasing opportunities for smaller farms. This production is rapidly growing. However, whether or not it will become a significant portion of US production remains to be determined.

There have been numerous studies done to examine the changes in population, loss of farmland and the related issues surrounding marginalisation of farmland in the United States. One study found that natural amenities are highly correlated with the change in rural population (McGranahan, 1999). The climate, water area and topography were used to construct a local index in this study. This index was used to examine the changes in rural populations. The impact of the different measures varies by

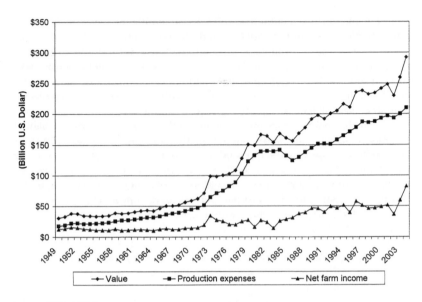

Source: USDA/ERS.

Figure 4.5   US value of agricultural production, total expenses, and net farm income

Source: USDA/ERS.

Figure 4.6   US net farm income as a percentage of gross farm income

area, but overall the natural amenities are a significant factor in whether or not a county increased in population and economic activity.

Another study examined the effects of several socio-economic characteristics on whether or not a county experienced population increases from 1990 to 2000 (McGranahan and Beale, 2002). Non-metropolitan counties classified as retirement or recreation counties, or ones with a predominance of federal land, experienced the most growth. These counties would all be counties that ranked high on the natural amenities scale. The counties classified as dependent on mining or farming showed the least growth. In fact, less than half the farming counties showed population growth during the 1990 to 2000 period. This is due to a variety of reasons, including the fact that the farming counties would tend to be low on the amenities scale, and that the changing nature of agricultural production has resulted in a loss in the number of farmers and supporting activities.

GOVERNMENT PROGRAMMES

There are a myriad of government programmes that affect the marginalisation of agriculture in the US. These programmes take many different forms and are provided by almost all levels of government in the US.

Farmland Protection

The largest category, in terms of number of programmes impacting upon marginalisation of agriculture, is farmland protection. The federal government, all 50 states and even some local jurisdictions have some form of farmland protection programmes. These programmes follow three basic approaches: zoning, regulatory or voluntary. Examples of the programmes include preferential tax treatment, easements, purchase or transfer of development rights, and the establishment of agricultural zones or other zoning regulations.

Nickerson and Barnard (2006) provide a detailed summary of the extent and nature of the current farmland protection policies and tools in the US. They note that preferential tax treatment, where agricultural land is taxed at a different rate, started in Maryland in 1956 and today all 50 states have some form of preferential tax treatment.

There are other forms of farmland protection used by many of the states. The use of the programme varies considerably by region of the country. For example, 19 states and 41 local jurisdictions in 11 states have some form of farmland protection programme involving the purchase of development rights to the property. However, just four states in the north-east US account for 76 per cent of the state level spending for purchasing development rights (Nickerson and Barnard, 2006).

The federal government also has some farmland protection programmes. The federal effort in this area started in 1981. The Agriculture and Food Act of 1981 required that the federal government evaluate the farmland impact of the various programmes that converted farmland to non-agricultural uses. The federal government became directly involved with farmland protection with the 1996 Farmland Protection Program. This involvement was expanded with the 2002 Farm Security and Rural Investment Act.

Hellerstein et al. (2002) published a study examining farmland protection laws and programmes in the United States. The purpose of their study was to determine how public preferences for rural amenities influenced the laws and programmes being enacted. Not surprisingly, they found that the influence of rural amenities varied considerably by region of the United States. Their study 'suggests that farmland preservation program emphases appear to depend on State-specific circumstances, including the amount of land already in parks, forests, and other conservation programs' (2002: iv). This study developed an extensive list of rural amenities or other outputs from farmland that served as a basis for the legislation they found. They condensed an expanded list of factors into major categories.

- A major impetus for farmland preservation was local and national food security. This affects various aspects of the marginalisation of agriculture and rural America. Many areas have started promoting local foods as a means of providing extra income for area farms and processors. There also is the appeal to people to be more environmentally conscientious and not consume foods that require so much fossil fuel to transport. Finally, there is the appeal to maintaining a diverse food supply for national and local security. Today 30 of the 52 states in the US have programmes for farmland protection focusing on food security (Nickerson and Barnard, 2006).
- Another frequently cited factor for farmland protection was the protection of environmental amenities. The Hellerstein study found a number of laws designed to protect the rural amenities themselves. Open space, the rural or agrarian character of the area, wildlife areas, natural areas and the overall aesthetics were important considerations in passing state and local laws and regulations. There are 29 states that have some form of farmland protection programmes designed to protect or enhance the environmental amenities of an area (Nickerson and Barnard, 2006).
- A third factor identified in the Hellerstein report was the overall protection of environmental services amenities. Citizens were concerned with pollution reduction, groundwater recharge, flood control, and water and air quality. Over 40 per cent – 23 – of all states have farmland protection programmes that are designed for the protection of the environment.

- Finally, a fourth, general category for farmland protection was orderly development. People were concerned about the orderly development of rural land to prevent sprawl. They favour the low-density physical space provided by farmland. Nickerson and Barnard (2006) report that 18 states have designed farmland protection programmes to encourage orderly growth and development of the area.

Other Programmes

Other programmes in addition to farmland preservation can aid rural communities. Some states and local areas have established rural development funds that provide grants or low-interest loans to businesses that locate in rural areas. Some communities provide tax and infrastructure support for businesses. There also are efforts to develop tourism and value-added industries, and other activities to help rural communities cope with the changes. Many local and state governments, as well as the federal government, are helping with programmes aimed at improving farmers' income and opportunities. Farmers' markets, community-supported agriculture programmes and the 'buy local' campaigns are examples of such activities.

Federal Programmes

The federal government has undertaken many programmes aimed at rural development. These programmes are carried out through many different federal agencies including the US Department of Agriculture. In many of the poorest rural communities, government transfer payments provide the majority of the income. Overall, government transfer payments made up 20 per cent of the income in the non-metropolitan counties.

With respect to the federal farm programmes, there are three major programmes designed to protect the natural environment. Each programme takes a different approach to preserving the environment.

The first of these is the Conservation Reserve Program (CRP) which was started 20 years ago under the 1985 Food Security Act. The CRP is a land retirement programme where the farmer is given a yearly payment to remove the land from production. Today there are more than 30 million acres enrolled in this programme. The CRP has been criticised for causing a decline in rural communities. The argument is that removing land from production leads to a decrease in the supporting activities, and this has led to the decline of communities in areas where there is a large percentage of acres in the CRP. However, recent US Department of Agriculture research 'indicates that, in aggregate, impacts have been limited. High CRP enrolment did not have a statistically significant adverse effect on population trends in farm counties across the United States' (Sullivan et al., 2004). This is an important finding because the use of land retirement programmes can be an important strategy to help improve environmental amenities. The CRP has been documented to improve water quality and wildlife habitat.

The Environmental Quality Improvement Program (EQIP) is another federal government programme. It provides cost share monies for improvements in the farming operation that enhance environmental quality. For example, the EQIP funds can be used to offset the costs of new manure storage facilities or to put in terraces. This has been a successful programme, but lack of funding has limited its usefulness.

The third major environmental programme is the Conservation Security Program (CSP), a new programme designed for working lands. Under the CSP, the farmer is paid for implementing practices that improve the environment. This is a voluntary programme and the farmer can choose to participate at three different levels. Each level requires more activities, but also offers a larger payment. The CSP was begun with the 2002 Farm Bill and has been available only on a limited basis to date.

These programmes show three alternative approaches that can be used to protect environmental amenities from possible damage from farming. Conservation programmes are a form of farmland protection. The programmes are designed primarily to protect the environment. Studies in the US have shown that protecting the environment, especially environmental amenities, is one of the best ways to ensure the orderly development of an area to prevent the marginalisation of agriculture (McGranahan, 1999; Hellerstein et al., 2002; Nickerson and Hellerstein, 2003).

A discussion on the impact of the federal government's programme on the marginalisation of agriculture would not be complete without mentioning the commodity programmes. The basic federal programmes affecting agriculture are contained in the farm bill. The US has a new farm bill every five years or so and this bill guides US agriculture policy. See Dimitri et al. (2005) for a discussion of the US farm bills and changes in US policy over time.

The reason for mentioning the commodity programmes is because, in spite of the fact there are many programmes at the federal level, the dominant programme type is the commodity programme. The commodity programmes represented 63 per cent of the estimated non-nutritional expenditure for the 2002 Farm Bill. By comparison the conservation programmes were 17 per cent and programmes for rural development were just 0.45 per cent of the estimated expenditure.

The current US commodity programmes are production based, especially for the commodities with programme support. The Economic Report of the President noted that: 'domestic support programs distort the price signals that farmers receive' (Economic Report of the President, 2006: 186). As such, these programmes create an environment where farmers' decisions are not based entirely on market signals, but also on government programmes. In 2005 the federal government spent approximately $20 billion on agricultural support payments. This would represent approximately 7 per cent of the total value of the US agricultural sector (Economic Report of the President, 2006).

The commodity programmes encourage farmers to keep land in production. In addition the programmes contribute to a loss of biodiversity and contribute to other environmental problems that have been associated with production agriculture. The commodity programmes have a significant influence on land values (Ryan et al., 2001). These programmes have been estimated to comprise 25 per cent of the land values nationwide and 45 per cent of land values in Iowa, a state receiving substantial government payments (Holste and Duffy, 2005; Ryan et al., 2001). Increased land values limit the ability of young people to enter farming and limit the access to land. Because of the size and magnitude of the government commodity programmes they are the dominant policy influencing the marginalisation of agriculture in the US.

It is debatable whether or not the commodity programmes contribute to or help stop the marginalisation of US agriculture. On the one hand the programmes help many farmers stay in business, and require conservation measures if the farmer is to remain eligible for commodity payments. But, on the other hand, they distort market signals and encourage intensive production of just a few commercial crops. This adds to the potential environmental damage from agriculture.

Regardless of the point of view regarding the direction of the impacts of the commodity programme on the marginalisation of agriculture, it is indisputable that they do have an impact. The commodity programmes represent the bulk of the non-nutritional spending on agriculture in the US. With a fixed amount of tax revenue, if the government is spending money on the commodity programmes, then the funds are not available for the other programmes that certainly address marginalisation of agriculture in the US.

CONCLUSIONS

One in five US counties depends on farming for a significant share of its income. But the face of US agriculture has been altered, and what will happen to these communities is not clear. The federal government has spent billions of dollars on farm programmes but these programmes have not helped maintain the farming-dependent counties.

Is marginalisation a problem for US agriculture? In terms of land abandonment, the answer is clearly no. There is very little, if any, farmland in the United States that is abandoned without another use. Farmland values in almost every region are at record high levels. There is, however, a considerable amount of farmland that is under pressure to change use for a variety of reasons. The amount of pressure depends to a large extent on the area and the natural amenities that are present.

Two major reasons exist for the marginalisation of US agriculture. The first is urbanisation. This is not necessarily just the growth of a city into the surrounding countryside. This is the movement of the people away from the cities into the countryside. In the state of Iowa for example, the number of people who live in the country but not on a farm now exceeds the number of people who live on a farm. People move to the countryside for a variety of reasons. Regardless of the reason, the shift does change the rural countryside. Land values increase as the demand increases. In addition, the demand for natural amenities increases. One of the major reasons people move to the country is for natural amenities, and they do not want to see these amenities destroyed or altered significantly.

The second major reason for the marginalisation of US agriculture is the changing nature of farms. On the one hand, as farm size increases there are fewer full-time farmers. As the number of farmers decreases, so does the number of businesses that the farmers support. On the other hand, there is a tremendous increase in the number of small, hobby-type farms. These farmers do not contribute as much to support the local economies, primarily because they purchase fewer agricultural inputs.

In the United States, the majority of farmers are small farmers who do not rely on farming for the bulk of their income. Over half of what are called farms in the United States either have retired operators or are lifestyle farms. These farms contribute to the rural countryside in a variety of ways, many of which are positive with respect to the preserving amenities. However, in some cases there is a natural tension between the part-time, lifestyle farmers and the full-time farmers. The lifestyle farmers are more concerned with the natural amenities, whereas the full-time farmers must also make a living from the farm.

US citizens enjoy a relatively cheap food supply. As such, for the most part, we only think about agriculture when something bad happens. Water pollution, worker problems, environmental problems and food safety scares are the only times when some people think about agriculture or the rural communities. The United States has a cheap food supply relative to the level of overall income, but there are hidden issues to be resolved. Current agricultural production techniques generate external costs that are not trivial. They have been conservatively estimated at between $5.7 billion and $16.9 billion annually (Tegtmeier and Duffy, 2004).

There is a movement to try and change people's attitudes toward agriculture and to think of agriculture in a broader context. We are recognising that agriculture provides a vast array of goods and services beyond simply producing food and fibre.

Some would argue that it is only the developed countries that can afford to worry about marginalisation of agriculture. They point out that the less-developed countries must rely on agriculture for food and fibre production first, and that other attributes or natural amenities come only after adequate production. However, this is a narrow point of view with respect to agriculture and its great potential. Agriculture can help with development, or it can hinder it. Developing countries must advance with a healthy agriculture as a base.

The alternative ways to view agriculture can have significant impacts. Multifunctional agriculture can help forge the bond between agriculture and the surrounding rural communities as a desirable place to live. Only time will tell whether this new approach helps alleviate the problems we are seeing in agriculture and rural communities.

NOTES

1. I would like to acknowledge the help and contribution of Dr Paul Lasley, Chair of the Iowa State University Sociology Department.

REFERENCES

Barnard, C. (2006), 'Farm real estate values', Agricultural Resources and Environmental Indicators, EIB-16, Washington, DC: US Department of Agriculture.

Dimitri, C., A. Effland and N. Conklin (2005), 'The 20th century transformation of US agriculture and farm policy', Economic Information Bulletin, EIB-3, Washington, DC: US Department of Agriculture.

Duffy, Michael (1994), 'Trends in farm profitability and regulations', Proceedings of Sixteenth Soil Seed Technology Conference, Iowa State University, pp. 85–94.

Duffy, M., D. Smith and J. Reutzel (2004), 'Farmland ownership and tenure in Iowa 1982–2002: a twenty year perspective', Iowa State University, Ext. Pub. PM1983.

Economic Report of the President (2006), Office of the President, US Government, February.

Fernandez-Cornejo, J. (2004), 'The seed industry in US agriculture: an exploration of data and information on crop seed markets, regulation, industry structure, and research development', Agricultural Information Bulletin, AIB-786, Washington DC: US Department of Agriculture.

Heffernan, W. and M. Hendrickson (2001), 'Consolidation in food retailing and dairy: implications for farmers and consumers in a global food system', National Farmers Union.

Hellerstein, D., C. Nickerson, J. Cooper, P. Feather, D. Gadsby, D. Mullarkey, A. Tegene and C. Barnard (2002), 'Farmland protection: the role of public preferences for rural amenities', Agricultural Economic Report, No. AER815, Washington, DC: US Department of Agriculture.

Holste, A. and M. Duffy (2005), 'Estimating returns to Iowa farmland', *Journal of Farm Managers and Rural Appraisers*, 68 (1), 102–9.

Hoppe, R.A., J.E. Perry and D. Banker (2000), 'ERS farm typology for a diverse agricultural sector', Agricultural Information Bulletin, AIB-759, Washington, DC: US Department of Agriculture.

King, J.L. (2001), 'Concentration and technology in agricultural input industries', Agricultural Information Bulletin, AIB-763, Washington, DC: US Department of Agriculture.

Lubowski, R.N., M. Vesterby, S. Bucholtz, A. Baez and M. Roberts (2006), 'Major uses of land in the United States, 2002', Economic Information Bulletin, EIB-14, Washington, DC: USDA/Economic Research Service.

McGranahan, D. (1999), 'Natural amenities drive rural population change', Agricultural Economic Report, AER781, Washington, DC: US Department of Agriculture.

McGranahan, D.A. and C.L. Beale (2002), 'Understanding rural population loss', *Rural America*, 17 (4), 1–10.

McGranahan, D. and P. Sullivan (2005), 'Farm programs, natural amenities, and rural development', *Amber Waves*, 3 (1), 28–35.

Nickerson, C. (2001), 'Smart growth: implications for agriculture in urban fringe areas', *Agricultural Outlook*, 24–7.

Nickerson, C. and C. Barnard (2006), 'Farmland protection programs', Agricultural Resources and Environmental Indicators, EIB-16, Washington, DC: US Department of Agriculture.

Nickerson, C.J. and D. Hellerstein (2003), 'Rural amenities: a key reason for farmland protection', *Amber Waves*, 1 (1), 8.

Reeder, R.J. and D.M. Brown (2005), 'Recreation, tourism, and rural well-being', Economic Research Report, ERR-7, Washington, DC: US Department of Agriculture.

Ryan, J., C. Barnard and R. Collendar (2001), 'Government payments to farmers contribute to rising land values', *Agricultural Outlook*, June–July, 22–6.

Stenberg, P.L. (2002), 'Communications and the internet in rural America', *Agricultural Outlook*, 292, 23–6.

Sullivan, P., D. Hellerstein, D. McGranahan and S. Vogel (2004), 'Farmland retirement's impact on rural growth, *Amber Waves*, 2 (6), 23–9.

Swine Odor Task Force (1995), 'Options for managing odor', North Carolina Agricultural Research Service, North Carolina State University.

Tegtmeier, E.M. and M.D. Duffy (2004), 'External costs of agricultural production in the United States', *Journal of Agricultural Sustainability*, 2 (1), 1–20.

Tyndall, John (2006). 'Shelterbelts and clean air pork', Odor and Manure Management, Iowa State University.

United States Census of Agriculture, United States Department of Agriculture/National Agricultural Statistics Service, various years.

USDA/ERS, 'Farm income data files', United States Department of Agriculture/Economic Research Service, http://www.ers.usda.gov/Data/ FarmIncome.

USDA/NASS (2004), '2002 Census of Agriculture, United States, Geografic Area Series', United States Department of Agriculture/National Agricultural Statistics Service, volume 1, Part 5, AC-02-A-51.

PART II

EFFORTS TO STRENGTHEN VIABILITY OF
AGRICULTURE

5. Northern European features and vulnerability to marginalisation: exploring indicators and modes of coping

Shivcharn S. Dhillion, Anna Martha Elgersma, Marja-Liisa Tapio-Biström and Hilkka Vihinen

Marginalisation is currently considered as a multidimensional process encompassing land abandonment and degradation, economic decay and deterioration in social and cultural dimensions. Depopulation, rationalisation in agriculture and forestry, and the establishment of protected areas for nature and landscape conservation are considered as the main changes in rural areas in Finland and Norway, and these changes point to the vulnerability to marginalisation. Vulnerability to marginalisation is often driven by unemployment, a lack of appropriate jobs, low income, no possibilities for education, opting for another lifestyle, the lack of a successor, remoteness and changes in policy (for example, Fosso, 1998, cited in Orderud, 1998; Almås, 1999; Nertsen et al., 1999; Grønbech, 2000). To cope with marginalisation several policy measures have been taken and interventions made; despite these measures, marginalisation continues. Throughout Norway's and Finland's long history, agriculture has structured the cultural landscape more than any other human activity. This is not limited to the cultivated land area, but also to non-agricultural land – the commons and forests.

What makes these northern nations different from most of the other European nations is their innate cold climate, sparse population and long distances. Marginalisation in essence is not a new phenomenon in the context of Finland and Norway although its conceptualisation, acceptance as an on-going process, discussion as a process in need for serious interventions, and its connection to globalisation and market forces are all issues relatively recently focused on. Combating marginalisation is a hefty task for governments, as a sustainable balance among local livelihoods, social services and socio-cultural components has to be attained. In this chapter we ask the following questions through two case studies, Mäntyharju in Finland and Tynset in Norway:

- What comprises the process of marginalisation in Norway and Finland, are there similarities?
- Do specific features of climate, sparse population and distances trigger marginalisation processes?
- What are the indicators for marginalisation in these cases?
- How is marginalisation coped with?

The chapter first presents a general description of land use changes and marginalisation in both countries, followed by the case studies.

A CASE STUDY APPROACH

Two case studies, Mäntyharju in Finland and Tynset in Norway, are used to study marginalisation. The case study areas reflect their innate cold climate, low rural populations, great expanse between settlements and long distances to urbanised areas. Thus both case study areas reflect the environmental handicaps for agriculture and remoteness in both countries in a European context. A set of indicators for marginalisation from previous national-level studies and from other studies on marginalisation was used. This set of indicators was complemented with indicators specifically designed for the case study areas. The number of indicators is limited to those most suitable in practice.

Municipality of Tynset in Norway (NUTS 5 level)

Tynset municipality is an inland municipality in the northern part of the county of Hedmark in East Norway. It is between 62 and 63 degrees of latitude and according to the European Union (EU)'s criteria of Less Favoured Areas (LFAs) it should be classified as an LFA area if Norway were a member of the EU. Population density is very low (2.9 persons per km^2), but not exceptionally low for rural Norway. The area has large distances among populated areas: the nearest centre with more than 5000 inhabitants is 150 km away. Tynset has 5463 inhabitants (1 January, 2004) and the area is 1806 km^2. The yearly temperature is minus 1 degree Celsius and the July temperature is 14.8 degree Celsius. Precipitation is low, 552 mm per year, and part falls as snow. The growing season is 100–140 days per year. Due to its climate the area is devoted to livestock farming and potato growing. Mountain farming is still practised. Mountains, shrubs, bare rock, peatland, large lichen areas and semi-natural and natural grasslands dominate the landscape and comprise 71 per cent of the total area. The utilised agricultural area (UAA) is 3.2 per cent and forest comprises 27 per cent, but forest production is low. The protected area is 18 per cent. The climatic conditions for agriculture are comparable with inland northern Norway (above the polar circle) and are classified as the worst class for production potential of the whole country (Mittenzwei et al., 2004).

Mäntyharju Municipality in Finland (NUTS 5)

Mäntyharju municipality is located in the south-western corner of the county of South Savo (NUTS 3). Mäntyharju has an area of 1210 km². Around 19 per cent of the area is water. The shoreline is 1520 km. The topography is so broken and irregular and the mainly moraine soils are so stony that suitable field areas are by necessity small and it is difficult to expand fields into larger units. Mäntyharju is a very rural municipality. The share of people employed in agriculture and forestry is 14.4 per cent of all employed persons; the national average is 4 per cent. In 1970 55 per cent of the employed people were still working in agriculture and forestry. This shows the late transition in Finland compared to other European countries. In 1970 there were 1117 farms and the average cultivated area was 6.34 hectares. At that time farms had on an average 43 hectares of forest. Forestry was still labour intensive at that time and was an important source of employment during the winter for the farmers. Due to technological development forestry is highly mechanised today with little labour input.

SUITABILITY OF LAND FOR AGRICULTURE AND LAND USE CHANGES IN FINLAND AND NORWAY

Large uniform forest areas, lakes, islands and peatlands characterise the Finnish landscape (Ministry of Agriculture and Forestry, 2003), while mountains, forests, fjords, lakes, bare rock, shrub, lichen and bog vegetations (Figure 5.1) predominate in the Norwegian landscape. Due to the huge extent of such areas combined with soils unsuitable or marginal for agriculture, cold climate, inaccessibility of land and in the case of Norway also steep slopes, the area suitable for agriculture is very small in both countries. In Finland it is only 7 per cent and in Norway 6 per cent of the total area. The inaccessibility of landforms and unsuitability of land for agriculture have caused agriculture to be concentrated in a few larger areas and in numerous very small areas scattered over the countries; distances between agricultural areas are large. Due to these features in Finland, agricultural land concentrates in the south and west of the country, where in some regions as much as 40 per cent of rural areas may be cultivated land. In the north, there are only small areas of agricultural land, mostly located in the sandy deposits of the rivers. Nearly all areas that can be cultivated have been cleared for agriculture in Finland. This is different in the case of Norway. In Norway agriculture is concentrated in the lowlands of southern, eastern and central Norway and 83 per cent of the area suitable for agriculture in these lowlands is UAA (Grønlund, 1989). In other areas the UAA is less, not only because of the limitations of the landform and related land cover and soil suitability, but mainly due to the restrictions of the climate. Only 21 per cent of the total land suitable for agriculture in the least favourable climate zone, northern Norway, is UAA (Grønlund, 1989). In total around 50 per cent of the area suitable for agriculture is used as UAA in Norway.

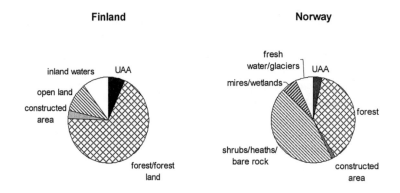

Sources: Statistics of Norway (2003), Ministry of Agriculture and Forestry (2003).

Figure 5.1   Main land cover types in Finland and Norway

Over time the area used for agriculture has changed and there is a trend that there is less need for land, but the process of change has not been similar for both countries. For example, for the period 1969–2001 in Finland the area used for agriculture and horticulture declined by 17 per cent, while for Norway it increased by 5 per cent. However, between 2000 and 2005 there has also been a decline in Norway (Fjellstad and Dramstad, 2005). In both countries there was an increase of land planted with cereals. In Norway the grassland area increased as well, but in Finland this decreased by 7 per cent. The abandoned land is covered by shrub and forest vegetation types after a few years.

Another significant change is that since the 1960s in Finland, nearly nine farms out of ten have given up cattle husbandry, while at the same time the overall cattle population has declined by over a third (MTT, 2005: 20). In Norway a significant change that occurred was the regionalisation of farm structures from the end of the 1950s. It was decided politically to increase cereal production on the best-suited soils and under the most-favourable climate conditions in the lowlands of eastern and central Norway; the remaining less-suitable lands were to be devoted to grassland and livestock production. The Nordic countries once had far more pastures than currently, and those that remain are cultivated pastures located further apart than before. The shift from pasture and grassland into cultivated arable land has in Norway led to an increase in erosion and in Finland to soil compaction. Mountain farming and grazing on the commons have always been part of agriculture in several areas of Norway, to increase fodder production, since that is low in marginal areas for agriculture with respect to climate and scarcity of UAA. In the past half century mountain farming and grazing in the commons has dropped significantly. The (semi-)natural grasslands and grazing patches in the commons are turning into shrub and forest vegetation communities due to the decline in grazing pressure.

A main reason for the decline in UAA, changes in land use and negative consequences for the environment is the shift from agriculture adapted to the marginal environmental conditions to agricultural systems not adapted to these conditions and the eventual closing down of the farms.

Features of Marginalisation in Finland and Norway

The concept of vulnerability of agriculture for marginalisation is very pertinent in both Finland and Norway since agriculture is totally policy dependent. Thus changes in policy will directly have an effect on the viability of agriculture. Due to the high latitude of both countries the growing period is very short and the July temperature is low, which results in a generally low level of production of crops per unit of area compared to many other European countries. The climate also implies a strong restriction in the variety of growing crops in most areas. Although the level of research and technology development is very high, the production level of crops remains low due to the climate. Thus production is maximised under the prevailing circumstances. At the same time the expenses are high due to the natural marginal circumstances for agriculture.

The marginality of the area for agriculture can be confirmed in that the whole area of Finland is classified as an LFA.[1] The most economically remunerative production line is dairy production, and there the productivity of animals is comparable to that of Central Europe, but due to higher production costs the economic result is dependent on the support mechanisms under the Common Agricultural Policy (CAP), and nationally financed support systems. The economic base of Finnish agriculture is slowly eroding with the reform process of EU agriculture and with the uncertainty connected to national aid in southern Finland. Thus the whole agricultural production system is very vulnerable to changes which seem constantly to worsen the economic environment of Finnish farms.

In Norway the marginal environmental conditions for agriculture are to a certain degree being 'demarginalised' by the very high national support to agriculture.[2] This situation makes agriculture very vulnerable to marginalisation when policy but also the farming systems are changing, in areas where agriculture is already at the edge of practice (that is climate-wise not feasible).

Both social and economic marginalisation of farms can be seen in the Finnish case to be results of a drastic structural change due to technological change and partial deregulation of the global food market, which excludes certain farms and certain areas from agriculture. The structure of Finnish agriculture has changed rapidly in the recent years. Before the EU membership in 1995 there were over 100 000 farms in Finland; ten years later there are about 71 000 left. The number of farms has fallen by more than 3 per cent a year. In livestock farming the decrease has been even more rapid, and the number of farms specialised in dairy husbandry has decreased by almost 7 per cent a year (MTT, 2005: 20).

Structural change includes the concentration of production both to fewer regions and to fewer farms; the change in production structure affects the economic viability of rural areas even more than the diminishing number of farms affects it. Production concentrates to areas where it is already strong, which indicates that the losing areas cannot compete under the present price and support level. The Finnish state has accelerated the structural change by directing investment aid and setting up young farmers with larger–than the average farms in the country. This is a deliberate choice, which is meant to make it easier for Finnish agriculture to adjust to the EU context (Ministry of Agriculture and Forestry, 2001: 53–54). But it is a choice based on a uniform strategic vision assuming that there is only one economically viable farm type, characterised by large area, high capital intensity and specialised production.

Despite the high level of support in Norway a similar decline in the number of farms by around 3 per cent annually takes place, like in Finland, particularly as small farms (Statistics Finland, 1960, 1970; National Board of Agriculture, 1983, 1992; Information Centre of the Ministry of Agriculture and Forestry, 2002). During the past five years dairy farming has decreased by 5 per cent annually while the size of the remaining dairy farms has increased. In Norway a goal of national agricultural policy has been to maintain agriculture across the whole country, but in reality the largest decline in farms in the past decades is in the north and remote areas (Norwegian Statistics, 2002). In Finland there is a tendency to maintain agriculture in climatically and topographically best-suited areas; however, the support policy seems to work in the opposite direction in the case of husbandry, at least based on the policy as formulated at present.

Thus, in both countries, with different policies, the general trend of a decline in the number of farms and the increase in the number of larger farms is quite similar. The decline of UAA may suggest that agriculture is in a process of marginalisation. In general the decline is caused by the fact that land is abandoned from agricultural use, and by soil sealing (Fjellstad and Dramstad, 2005). Another feature is the urbanisation process. The concentration of population in Finland to southern and coastal areas has left large areas of the eastern and northern parts of Finland extremely sparsely populated. In Norway a similar trend is taking place; there is an increasing development of out-migration from northern and central Norway to the southern part of Norway and to larger centres (Elgersma et al., 2004). In Norway the situation is such that over 50 per cent of the municipalities have less than 3200 inhabitants and cannot offer their inhabitants the necessary services any more (NOU, 2004).

To summarise the main features of marginalisation are: abandonment of agricultural land, particularly in Finland; decline in farms, in particular in the more sparsely populated and remote areas; concentration of agriculture in better-suited areas for agriculture; and depopulation of rural areas. As the features of marginalisation are not similar within and among the countries, marginalisation is studied in two local cases.

Case Studies in Mäntyharju and Tynset – Disentangling Marginalisation

Land use changes and marginalisation: climate, land quality and policy driven

The size of the UAA in both case study areas is strongly limited due to irregular terrain conditions, and thin and stony soils. In the case of Tynset this is also due to steep slopes. Thus the area suitable for agriculture in both areas is small. The current UAA is only 4 per cent in Mäntyharju, which is nearly half of the country level, and is 3.2 per cent in Tynset reflecting the national average (Table 5.1). In Mäntyharju the UAA has declined by 47 per cent since the 1970s and explains the even smaller area used for agriculture at the country level. This decline is clearly a sign of marginalisation. This is different for Tynset where the UAA increased by 26 per cent in the same period (Tables 5.1 and 5.2a and 5.2b). It is clear that in Tynset no significant agricultural land abandonment has taken place as it has in Mäntyharju. From 2002 to 2003 there was a small decline in the UAA in Tynset. In fact, in Tynset nearly all land suitable for agriculture is UAA.

Table 5.1   Main indicators for marginalisation in the case study areas

Category	Indicator	Mäntyharju (Finland)	Finland	Tynset, Norway	Norway
Land use	Share of UAA in total area (%)	4	7	3.2 (1999)	3.2 (1999)
	Share of old forest (%)			38	32.9
	Abandoned UAA (1969–2000) (%)	47	17.4	neglectable	neglectable
	Protected area (%)		9 (2003)	18	12
	Grazing pressure commons (sheep) (Decline in % 10 years)	not applicable	not applicable	-1	4.3
	Soil quality	stony moraines	varied, from clay to moraine	no restric- tions	low fertility stoniness, shallow soil, erosion risk
Climate	Growing season (days)	165	100–170	100–140	100–220
Population	Population density (persons per km2)	7	15	2.9	15
	Population change (%)	-30.7 (1970–2002)	13.2 (1970–2002)	6.7	17
	Share of population over 65 years (%) (2002)	23.4	15.3	15.9	13.2
Socio- economic	Average farm size (ha)	16.7 (2002)	29.9 (2002)	17.4 (1999)	14.7 (1999)
	Decline farms; % yearly (past ten years)	3	3	3	3
	Unemployment rate (%) (2001)	14	12	1.9	3.2
	Off-farm income/total income per farm (%)	25 (2001)	28 (2001)	38 (1999)	47 (1999)
	Net farm income per capita (€)	12 958	17 997	17 489	19 335
	Share of farm support in net farm income)(%)		165 (2002)	118	110

Sources: Norwegian figures are from Statistics of Norway (2003b) and Tynset municipality (2003). Finnish figures are from Information Centre of the Ministry of Agriculture and Forestry (2002), National Board of Agriculture (1970), Finnish Forest Research Institute (2003), Rikkinen (1992) and Mäntyharju municipality (2005).

Table 5.2a   Land use per category of used agricultural land (ha),
*   Tynset municipality*

Land use	1969	1979	1990	1995	2000	2002	2003	Change 1969–2003 ha	%
Total UAA	4455	4681	5144	5497	5595	5794	5630	1175	26
Meadow and pasture	4032	4265	4684	5069	5175	5398	5252	1220	30
Potatoes	18	16	18	33	75	103	73	55	306
Grain (barley)	19	43	43	10	61	62	90	71	374
Other	385	357	399	385	284	231	215	-170	-44

Source: Statistics of Norway (2003b).

Table 5.2b   The farmland use in Mäntyharju, 1969–2000 (in ha)

Year	Agricultural and horticultural land, total	Rented arable land	Arable land under cultivation	Grazing (rough) and pasture (meadows)	Forest land	Other land	Total
1969	7 141		7 141	544	46 553	3 649	57 889
1980	5 932	253	5 501	512	36 888	3 058	46 391
1990	5 206	514	4 691	345	32 493	2 536	40 580
2000	3 821	1 174	3 749	69	16 307	1 251	21 442

Sources: Information Centre of the Ministry of Agriculture and Forestry (2002), National
         Board of Agriculture (1970, 1992, 1983).

Similar to the country level, marginalisation of agriculture is evident as both areas have undergone a strong decline in farming as an occupation. In both areas the decline in the number of agricultural holdings was 3 per cent yearly between 1995 and 2005, following the national trends.

Simultaneously a tremendous structural change with respect to increasing production in agriculture took place. In Mäntyharju in 1980, 84 per cent of all farms were under 10 hectares; this share has declined to 37 per cent, but this percentage is still high. In Tynset in 1979, 57 per cent of the farms were less than 10 hectares and this reduced to 24 per cent in 1999. An explanation for the differences in farm size is that the land parcels in Mäntyharju are very much spread, isolated and small (average 1.6 hectares) and cannot increase in size due to the irregular terrain conditions. In Tynset a large number of the land parcels are situated in the bottom of broad river valleys and on natural terraces. The parcels in the bottom of the river valleys can be consolidated and are therefore larger than in Mäntyharju (2.7 hectares) and are by large easily accessible. The advantages of the consolidation of parcels in Tynset compared with Mäntyharju points to a reason why land is not abandoned as it is in Mäntyharju. Larger parcels and bordering parcels are more effectively managed than small and isolated fields. The UAA per farm is in both areas fairly similar (Table 5.1): 18.6 hectares in Mäntyharju (2003) and 17.4 hectares in Tynset (1999). The terrain difficulties for enlarging farms in Mäntyharju are reflected in that the average farm size in Finland is much larger, 31 hectares in 2003. In Tynset farm size is a little larger than the national average. This may suggest that proportionally, agriculture in Mäntyharju is more vulnerable to marginalisation than in Tynset.

The process of land use change and marginalisation is evident, but different for Mäntyharju and Tynset. However, while a large part of the fields have been abandoned in Mäntyharju, some new fields have been cleared in both areas. In Tynset these new fields are often of lower soil quality. Land use has also changed in other respects. In the early 1970s there were few rented fields in Tynset and none in Mäntyharju, while in 2000 around one-third of the cultivated land was rented in both areas. The meadows in Mäntyharju have almost all disappeared, whereas this has not occurred in Tynset (Table 5.2a).

In both areas agriculture is combined with forestry. The forest area owned by farmers[3] in Mäntyharju has diminished by around 65 per cent between 1969 and 2000 due to rural exodus and related inheritance, and 40 to 50 per cent of the forest owners live elsewhere. However, forest revenue is still an important part of the farming economy in Mäntyharju. In Tynset many farmers who close down the farm remain living in the area, and thus the majority of the forest owners still live in the area. But low timber prices combined with high exploitation costs, low productive forest due to climate, and poor soil and little time for forestry work have caused some decline in forestry. Although one may interpret marginalisation of forestry in Tynset to be indicated by the high share of old forest, which is 38 per cent and higher than at the country level (Table 5.1), maintenance and increase in old forest over the years has been the forest conservation strategy and policy. The policy encourages old forest maintenance and selective use.

The agricultural land use development as related to marginalisation is by no means straightforward. In the first years of the twenty-first century there was a slowdown in the decline of field area in Mäntyharju, while decline in the UAA in Tynset started in the twenty-first century. In Mäntyharju there is a renewed interest in keeping fields in some sort of cultivation now, when the EU subsidies are increasingly attached to the area cultivated. This has led to a situation where there are at the same time some farmers who clear new land, while others reforest. As one interviewed person said: 'We are clearing forest for new fields while our neighbour on the other side of the road is planting trees'. Farmers investing in increased milk production also need more land, for example to spread manure in Tynset, but most of the increased area is obtained by renting. However, at the moment the interest in investing in dairy production is very low in Mäntyharju. This naturally leads to a decreased need for fields and, particularly, fields further away from the farm houses are abandoned.

In both countries reforestation support was given to increase the forest area and, in the case of Finland, to decrease overproduction in agriculture. After 1995 this support system was terminated. In Mäntyharju around 750 hectares of fields were reforested between 1970 and 1995. The rest of the fields which have been abandoned from cultivation, altogether 2570 hectares, have probably started to grow shrub and later forest through natural processes. In Tynset the reforestation support was little used as forest production is low and forestry generally is not an important source of income for farmers, whereas this is the case in Mäntyharju as forests are highly productive.

To cope with the marginal character of agriculture in both areas due to climate, and in the Finnish situation also due to soil conditions, dairy farming has traditionally been the main farming system in both areas as arable farming is not viable. The low production level of fodder due to climate and scarcity of the UAA in Tynset necessitated in the past the use of the large area of the commons for grazing and mountain farming. Due to the intensification and effectivisation of agriculture the economical viability of grazing on the commons and mountain farming has declined, but is still high compared to other areas (Table 5.1).

In Mäntyharju there has been a shift from milk production to grain production since Finland joined the EU. The reason for shifting from dairy production to arable production is policy driven. According to the Act of Accession (Article 142), Finland is allowed to pay national northern aid north of the 62nd parallel and in adjacent areas, that is support area C. The most important aid measures are northern milk production aid (€202 million in 2004) and northern aid based on livestock units (€105 million in 2004). Mäntyharju was unfortunate in the sense, that it was left just on the southern side of the northern aid border, even though the natural production conditions are at least as poor as in its neighbouring municipalities. As a result, a dairy farmer in Mäntyharju, whose farm is classified as the national aid region B, earns on average €2000 less in subsidy per year than his colleague in the neighbouring municipality. Because subsidies form a decisive part of farmers' income, many dairy farmers in Mäntyharju have lost motivation, given up the cows and continue arable farming, trying to find other income sources. In the Finnish case, production decisions depend heavily on the design of subsidies, since they form 40 per cent of farmers' income. As the administratively set producer prices in EU are very low for Finnish production costs all over the country, the conditions for support are crucial. In this case, the subsidy is likely to shift dairy production to the northern side of the national subsidy border – it is not a matter of production moving to areas with better production conditions, but to areas more profitable in the sense of subsidy maximisation.

There are also some options for other production lines in agriculture in Mäntyharju. While pig production is also declining, since only very large units are profitable today, specialisation in horses is a growing trend. Considerable amounts of development resources have been invested to improve the potential for equine husbandry in Mäntyharju. Vegetable and berry production has an established share of farms in Mäntyharju, the number has been 15 farms between 1995 and 2005. However, Poland joining the EU has created fears of too intense a competition to the detriment of berry production in Mäntyharju, and Finland in general. To compensate for the giving up of dairy production in Mäntyharju, many farmers have chosen a pluriactivity strategy. There is shift from full-time professional farming to pluriactivity with other income-generating activities on- or off-farm and/or wage employment.

In Tynset there has also been a shift away from milk production to other production lines since the mid 1990s, because off-farm employment is becoming a more and more important source of income and dairy farming is too time-consuming to be combined with this. Two processes that are taking place now are: (1) a shift from dairy farming to beef cattle and sheep farming and intensification of dairy farming; and (2) keeping horses and growing potatoes is increasing. The soil is very suitable for growing potatoes adapted to a cold climate.

The Interactions between Agriculture, other Land Uses, Landscape and Marginalisation

The environmental restrictions due to the harsh climate and small area suitable as farmland has in the past given rise to farm structures adapted to these restrictions. In both areas agriculture has always been practised in an intimate, functional relationship with the remaining land areas.

The changes in agricultural land use are part of the restructuring process of rural areas. The technological change also affects other major land users, including forestry, mountain farming and grazing in the commons. The economic importance of forestry as part of the farm income has decreased with technological change, although it is still significant. Formerly, work in one's own forests or somebody else's forests was, during the winter period, an important income source for small farmers in both areas. Now manual labour has been replaced by huge machines which are owned and run by machine contractors. As the income from forestry for farmers is much higher in Mäntyharju than in Tynset, the consequences of the mechanisation are also larger. In Mäntyharju the forest income decreased from 35 per cent of the total farm income 1990 to 29 per cent in 2002, but this is still high compared to the national average of 11 per cent in the year 2002 (Mäntyharju municipality).

The technological change and increased effectiveness of forest production has also meant similar types of changes in the forest quality as is the case with agriculture in Mäntyharju. The large, uniform, cultivated forests with large, uniform, stands contribute to the loss of biodiversity and create a monotonous landscape. In Tynset mechanisation in forestry has not led to large, uniform, cultivated forest stands. The point here is that a decline in forest clearings took place and caused a homogenisation of the age distribution classes of the forest stands, which impacts upon biodiversity negatively and homogenises the landscape. In Tynset, due to the cold climate, the forest is better suited for quality production than for bulk production, but this potential is little used. The decline in mountain farming and grazing in the commons has increased the closing of forest and the disappearance of (semi-)natural grasslands. Additionally as many mountain farms are in status of a decline in cultural heritage is observed and referred to by many.

Since in both areas the fields have always represented a very small share of the total land, any reduction in field area has to be considered as a serious threat to the openness of the landscape and the decline of forest edges, which often have high value for biodiversity. Also, the cultural landscape changes as both permanent, uninhabited buildings and their surroundings fall into decay. For instance, in Tynset over 30 per cent of the farms have no permanent residents. Cultivation is concentrated as much as is geographically possible around the farmhouses. The so-called 'thrown around' fields, meaning small fields, maybe as small as 0.10 hectares situated some distance from the farmhouse in the middle of the forest, are increasingly abandoned in Mäntyharju. The multifunctional role of agriculture contributing to the agricultural landscape and biodiversity has declined in both areas. In both areas little attempt has been made to secure the future of the most valuable landscapes.

A significant difference between Mäntyharju and Tynset is the change in the mode of production in Mäntyharju. The change in production structure of the farms from dairy production to grain cultivation has the greatest impact on biodiversity. Animal production is connected to more varied land use practices, with pastures and the production of hay. Grain production means a more monotonous ecological environment and a decrease in the number of species, many of which are dependent on meadows and pastures and some of which are under threat of extinction (Hänninen-Valjakka, 1998: 17–19).

While fields will in the future be held in cultivation because of the subsidies tied to the land and the connection between the product and the income of the farmer is lost, farming loses both its economic and its moral ground.

Features of land use change linked to marginalisation are associated with settlement patterns. There is a trend of concentration of the population in municipality centres. Tynset as well as Mäntyharju show this trend.

Pluriactivity and Multifunctionality

With the existing farming structure in both areas, it is not surprising that other gainful activities are important. There are various sorts of other industries besides agriculture in 37 per cent of the farms in Mäntyharju. This can be compared to the national average of 27 per cent (MTT, 2004: 16). In Tynset this figure is 50 per cent (Table 5.3) while the national level is 38 per cent. Pluriactivity might be very vulnerable to competition as there is no need for similar services at a large scale in areas with very low population density. Despite of the importance of pluriactivity on the farms for enhancing incomes, in Tynset the income from pluriactivity is only 8 per cent of total income for dairy farming households and 9 per cent for sheep farming households.

Table 5.3   Supplementary activities on the farm in Tynset (1999)

Supplementary industries	Holdings (n)	Holdings (share of total number of holdings, %)
All activities	161	49.5
Contracting work with tractor	90	27.9
Camping site, cabin renting	11	3.4
Renting out buildings on the farm	13	4.0
Other	106	32.8

An off-farm job is in Tynset the main way to maintain farming. Another way to cope with marginalisation is developing networking among farmers to start projects together, and it appears that networking is becoming an important way to maintain farming in the area. For example in Tynset, in the hamlet of Tylldalen, a number of farmers (up to 12) have formed a group collaborating on a special veal production system.

In Mäntyharju the informants had different opinions about the multifunctional use of fields. The issue was addressed with a question: Are fields as scenery worth money? Two of the interviewed persons were totally negative to the idea that farmers would be paid for cultivating their fields in order to preserve the scenery. 'It feels bad, if fields are cultivated only because of the scenery, then there would be no need for the professional competence of the farmer, this would be mental violence!' As for the economics of it, the support should make cultivation of fields comparable to the economic returns of forestry.

In Mäntyharju the interviewed persons were unanimous. The shift to farming that has no connection to the price of the product will in the long run be the end of agriculture in the area. Participants also lamented the fact that there is a tendency for the cultivated areas to become more monotonous. This is very much due to the policy, as milk production is decreasing and grain cultivation is increasing. The value of the grain yield is so little compared to the production costs that in difficult years like 2004, when it was very wet, it was tempting to leave the crop in the field. In Tynset, where such changes did not take place, the great fear for marginalisation of agriculture is possibly the decrease of the very significant state support.

Other Land Uses

Despite the 'negative' development of the multifunctionality of agriculture, the areas still have a high potential for development of new types of land uses, due to their very low population density, clean air, nature and cultural heritage. Both areas have excellent opportunities for outdoor leisure recreation. In Tynset a start has been made with developing such new land uses. The focus is on the specificity of the area compared with other areas, that is mountain farming. Mountain farming, cultural heritage, sleigh riding in wintertime and the opportunities for recreation offered by the national park and other protected areas, are the basis for development of such new land uses. Farmers are actively involved in these developments for two reasons: they own a large part of the land involved, and the activities to be developed offer new possibilities for pluriactivity. The key issue for success is the development of infrastructure for such new land uses. The obstacles are that Tynset is not traditionally in a tourist area in Norway, it is far from urbanised areas and the summer season is very short. The low population of the whole region is another obstacle, as few local people can participate in local recreation. The development of cottage tourism, as in other areas in Norway, has not taken place. Presumably distances to larger urbanised areas are too great. This is different for Mäntyharju.

A specific feature in Mäntyharju is no doubt the great number of summer cottages located in the municipality, almost 4500. There is a long tradition of summer villas in Mäntyharju: the oldest ones in the centre are over 100 years old. The natural conditions for summer cottages are excellent, with the great number of lakes. The summer guests double the population of Mäntyharju for some 86 days a year, which is the average time they stay in their cottages in Mäntyharju annually. The economic impact of summer guests is clearly visible in the Mäntyharju municipality centre. The number and type of shops, for example two shops specialising in interior design, is exceptional for a small municipality like Mäntyharju. While national rural tourism is important in Mäntyharju, in order to make it into a strong industry in the area, the summer residents should be encouraged to consume services that have a proper price, and to buy more local products. Local business development efforts are needed.

Support and Programmes

In Tynset the high level of support as part of net farm income (see Table 5.1) and the high state and county subsidies in maintaining public services in the municipality contribute to the fact that marginalisation is not felt as being a problem. It would become problematic if subsidies were to decline. Maintaining the services in the municipality are of crucial importance for maintaining agriculture as it offers possibilities for off-farm jobs and keeps unemployment at a low level (Table 5.1). The acreage-based and headage payments, the most important direct payment categories, strongly contribute to keeping the land in cultivation. The marginality of Tynset for a wide range of agricultural purposes is also due to the fact that Tynset falls into zone 5 of the acreage-based support (Table 5.4), which implies that growing coarse fodder and potatoes is promoted. The introduction of acreage-based payments in 1989 also aimed at keeping particularly small farmers in business. This, however, has not occurred in Tynset and elsewhere in the country: the payments have, however, contributed to maintaining the openness of the cultural landscape as little land is abandoned in Tynset. The recently initiated national ten-year food and biofuel programmes and support for organic farming have been little used until now. Compared to other regions in the country, mountain farming is still much more commonly practised and is seen as one of the specialities (the products) of the area and a basis upon which to develop new land uses. Organic farming is also stimulated but little practised. The problem is that sale of local labelled or organic products is difficult since there are few consumers in the area, the distance to larger markets is large, and a national marketing infrastructure is lacking.

In Finland as an EU nation there are several programmes for enhancing rural and regional development partly financed by the EU. Mäntyharju belongs to the Objective One programme of eastern Finland and to the Objective Six area under the former programme period. There are several development projects in the Mäntyharju area, for example aiming at improving dairy cattle health and the genetic base, diversification of farms, processing of farm and natural products, and the development of rural tourism.

Finland has mainstreamed LEADER-type local action groups (LAGs) to the whole country. Mäntyharju belongs to the Veej'jakaja LAG, which is financed nationally, but operates along the same principles as LEADER+programme groups. Veej'jakaja has projects with several other associations, providing strong support to the development of rural industries. The LAG activities have encouraged cooperation between villages and revitalised the old talkoo heritage, which means gatherings for voluntary work with a well-defined goal. There is development dynamics in Mäntyharju to a great extent. The long-term village movement, and the local action groups with their economic resources which were established after EU membership, have built a social capital which is clearly a force to be recognised.

Driving Forces of Marginalisation

The fundamental handicap of agriculture in Finland and Norway is the cold climate, and in Norway also the inaccessibility of the terrain and steep slopes. Climatic conditions determine the location of crop production and the production level. There are severe restrictions on the variety of cultivable crops that can be grown related to climate and physical conditions.

Other driving forces are changes in the economic and social environment, globalisation of agricultural trade, technical development, policy changes adapting to the new socio-economic and ecological circumstances, and also the decreased attractiveness of farming as an occupation, and social priorities. Net farm income compared to income from other occupations is low. This is considered as being the main driving force for closing farms. Partly this is due to the very high costs, as costs are in marginal areas much higher than in better-suited areas (Romstad, 2004).

A specific reason for marginalisation in Finland and Norway is the low density of the rural populations and large distances.[4] These together contribute to the problematic market situation. The markets are far away and relatively small, which makes it difficult to develop local and regional solutions based on production for local markets.

Coping with Marginalisation in Finland and Norway

In Norway the main way of coping with marginalisation of agriculture is through market price support and direct support. Market price support is the price difference between the higher prices the farmers get for their products, and the lower world market prices. This is made possible due to high import restrictions. Due to the market price support, domestic production is greater than it otherwise would have been (Rogstad, 2003: 10). To maintain agriculture in marginal areas, several direct support schemes are partially differentiated in terms of production, geographical region and farm size. An example of this is the acreage and landscape support scheme of 2005 shown in Table 5.4. Since 2005, a flat rate of support is given for all UAA, whereas there is a differentiation for supplementary support for various crops. Small farms are most favoured by this scheme as the first 20 or 3 hectares receive the highest support, in particular for coarse fodder and vegetables, berries and fruits. The acreage and landscape scheme was introduced in 1989 and may presumably explain the increase of UAA as farmers more precisely report their entire land acreage to the authorities; this is particularly the case for grassland and pasture. The goal of the introduction of the flat rate of support per hectare is to preserve the open agricultural land and cultural landscape. Since the 1990s, there has been a shift from price support to production support to preserve the environment and the agricultural landscape.

Table 5.4   Acreage and landscape support (AC payments) per ha in € (1€ = 8.0 NOK) for 2005

	Area size per farm (ha)	Zone 1	2	3	4	5	6	7
Agricultural area in use	Total area	234						
Supplementary support:								
Coarse fodder	0–20	93	0	123		238	282	320
Coarse fodder	>20	62	0	62		62	62	62
Grain	Total area	120	182		290			
Potatoes	Total area	62					1125	
Vegetables	0–3	500					1875	
Fruits and berries	0–3	625					1375	

Note: Zone 1: according to climate best suited for agriculture; zone 5–7 worst-suited; zones 6 and 7 are for northern Norway.

Source: http://www.slf.dep.no.

For Finland membership of the EU meant a dramatic change in the economic environment of farming. When Finland joined the EU in 1995, institutional and market conditions changed overnight. On average, producer prices decreased by 40 per cent when the price subsidies were wiped out. Without massive support from the EU and national sources, Finnish farms would have gone bankrupt under the prevailing EU price level.

The policy design also changed dramatically. The subsidies, which were earlier predominantly included in producer prices, were mainly transformed into direct income support. Regional subsidies and small farmers' support were abolished, which has had regional and income distribution effects accordingly. The unfavourable natural conditions make the role of support in the income of agriculture much greater in Finland than in other parts of the EU – agricultural support accounts for 44 per cent of the total output of agriculture and horticulture (MTT, 2004).

The CAP support system is not suited to Finnish conditions (or has negative consequences in a country like Finland). Because of Finland's geographic location and climate, CAP support largely based on yields per hectare, which occupies a central position in the EU policy, represents only 25 per cent of all support paid in Finland. Seventy-five per cent of agricultural support consists of three other measures: compensatory allowances paid for the natural handicap, the agri-environmental scheme and national aid. In the membership negotiations it was agreed that LFA support would cover 85 per cent of the arable land area, but since June 2000 the LFA covers the whole cultivated area in Finland – the first EU country where LFA support may be paid in the whole country. Also the coverage of agri-environmental support is a record high within the EU: at the end of 2002, the environmental support covered 94 per cent of Finnish farms and 98 per cent of the arable areas (MTT, 2003). LFA and environmental support are co-financed, whereas the national aid – northern aid, national aid for southern Finland, national supplement to the environmental support and certain other measures which aim at securing the preconditions for Finnish agriculture in the different sectors and regions – is paid completely from the national funds.

Finland suffers from the fact that the main CAP support in the form of export subsidies and direct payments depends on yields, and rewards countries and farms with the best competitiveness, best natural conditions and farm structure. No matter how much Finnish farms invest, they will never be able to bridge the gap caused by the climatic and topographic disadvantages.

Studying the effects of the structural policy and the CAP of the EU on remote, poor rural areas with diversified and agriculturally dependent economic structures, Efstratoglou et al. (1999) showed that the total economic effects of traditional CAP support in the local economies are lower than those of investments promoted by the structural policy.

The main logic that the CAP still follows inevitably causes marginalisation of agriculture in Finland. Finland uses all possible means to support farmers' incomes, even with measures which have originally been formulated for other aims (LFA and agri-environmental measures). The volume and breadth of the so-called second pillar measures are far from sufficient to counteract the marginalising effect of the logic in the first pillar measures.

There are two possible farming strategies to cope with marginalisation: either by increasing farm size or by changing to more productive special crops. Enlarging farm is to a large extent done by renting land, and 40 per cent of the total cultivated area in Finland is leased, while for Norway this was 31 per cent in 1999. Support and initiating new types of production are core to increasing the viability of agriculture. In Norway ten-year national subsidised programmes such as the biofuel and food programme have been put forward. Also in Finland the production of biofuels in the fields is considered to be a serious option for farms.

The other option for the farms to survive is pluriactivity. Apart from traditional income from farming and forestry, income is generated from other businesses on- or off-farm, or from off-farm jobs. In Norway the net income from agriculture and horticulture was only 27.5 per cent of the total farmer's and spouse's income in 2001. A small part of the total income is from other activities on the farm (8 per cent); most of these activities are machine contracting. The largest part of income on the farms in Norway comes from an off-farm job (48 per cent). Having an off-farm job is called 'a survival strategy of agriculture' (Blekesaune, 1996).

In Finland 27 per cent of the farms practiced other industry besides farming in the year 2000. Most of these farms were located in the south-west and the west. The farms engage in various kinds of activities: most common are machine contracting (41 per cent of farms), tourism, other services, and food and wood processing. Entrepreneurial activities on the farms are usually quite small: in 42 per cent of the farms their annual turnover was less than €8250 (MTT, 2004: 16–17).

The importance of other gainful activities than agriculture as the income source of farmers and spouses of farmers has grown remarkably in both countries. This means that there are now more part-time farmers and the share of full-time farmers has clearly decreased. Fifty-six per cent of farmers could be classified as part-time farmers, and the rest (44 per cent) as full-time farmers in 1999 in Finland, from private farmers under 65 years old and with at least 2 hectares of agricultural land (Peltola, 2000). In 1973 on 65 per cent of all farms, the share of agricultural net income of farmer's and spouse of farmer's total income was 75–100 per cent. In 2000 only 40 per cent of farms had that income profile (Vihinen et al., 2005: 51).

Generally, the importance of other gainful activities has been, during the past two decades, lowest in eastern Finland, while it has been highest in the south and west. There is a natural explanation for this, related to the fact that other gainful activities (paid jobs outside the farm, or markets for non-agricultural products or services) are much better and more plentiful in the south and west of the country, which are more densely populated, than in the north and east (Vihinen et al., 2005: 52). In Norway a similar trend is occurring: the percentage of full-time farmers is highest in the more sparsely populated areas. However, this can also be explained by the fact that dairy farming is concentrated in the sparsely populated areas; this is partly the explanation in Finland also. Animal husbandry is more time-consuming than other kinds of production, which means that less time remains for other activities on husbandry-dominated farms.

CONCLUSIONS

Cold climate, large distances and sparse population are conditions which will not change in Finland and Norway. These conditions are serious handicaps for agriculture and rural development. Agriculture is not and will not be competitive for any type of production without support. Thus agriculture is totally support dependent in both countries.

The large distances and sparse population are also a severe handicap for the enhancement of pluriactivity. Since the possible customers are few and far away it is not easy to start value-adding or other activity on farms. For example finding customers for different specific food products processed in small amounts on- or off-farm in rural areas has high transaction costs, since potential buyers are far away from the producer and from each other, and also the retail shops are organised in large chains which procure centrally and are not interested in purchasing small quantities.

In the case of Finland, and Mäntyharju in particular, we have to underline the fact that most of the subsidies farmers get are either LFA, environmental measures or national support based on specific geographic difficulties. The rationale behind these measures is connected to regional balance and multifunctionality. However, they seem not to be able to combat the marginalisation process.

In Norway support measures for specific geographic difficulties, and acreage support to maintain farming in marginal areas and to enhance multifunctionality, have prevented the total acreage of agricultural land from being significantly reduced. However, marginalisation of agriculture as an occupation continues.

The change in growing crops not adapted to the local environmental conditions in marginal areas, due to changes in support schemes as in Mäntyharju, makes areas marginal for agriculture even more vulnerable to marginalisation as production reduces rapidly. The acreage support scheme in Norway counteracts such developments by stimulating the growth of crops that are suitable for such a marginal area, as is the case in Tynset.

In both areas development of new land uses to combat marginalisation is partly based on traditional farming systems (Tynset), traditional pluriactivity, the specialities of the areas, and in the case of Mäntyharju also on growing new crops like caraway. These to some degree appear to point to potential viable ways of combating marginalisation. In both countries there is an increasing interest in developing these lines further, although the momentum is slow and requires perks. The availability of off-farm jobs is crucial for combating marginalisation. However, in sparsely populated areas and large distances to centres finding an off-farm job may be a problem. Again, the specialisation of areas and specialty product development is a viable aspect to build upon.

NOTES

1. As Norway is not a member of the EU, no area is designated as a LFA.
2. The support to agriculture in 2001 expressed as percentage PSE (OECD's Producer Support Estimate) was 66 per cent, whereas this was 34 per cent for the EU-15 (Rogstad, 2004).
3. In both areas farming is combined with forestry and the largest share of the forest area is private: 78 per cent in Mäntyharju and 86 per cent in Tynset.
4. Both countries have a population density of 15 persons per square kilometre but in Norway in rural areas, which is the largest part of the country, it is less than three persons per square kilometre.

REFERENCES

Almås, R. (1999), 'Rural development: a Norwegian perspective', Trondheim: Centre for Rural Research, NTNU, Report 9/99.

Blekesaune, A. (1996), 'Family farming in Norway: An analysis of structural changes within farm households between 1975 and 1990', Doctoral thesis, Department of Sociology and Political Science, University of Trondheim, Report 6/96.

Efstratoglou, S., D. Psaltopoulos and J. Kola (1999), 'Structural policy effects in remote rural areas lagging behind in development (STREFF): executive summary', Athens.

Elgersma, A.M., M.A. Støen and S.S. Dhillion (2004), 'Status of marginalization in Norway: Agriculture and Land Use', EUROLAN report 2005/6, (http://www.umb.no/ina/forskning/eurolan/index_e.htm).

Finnish Forest Research Institute (2003), 'Finnish Statistical Yearbook of Forestry 2003', Helsinki.
Fjellstad, W.J. and W.E. Dramstad (2005), '3Q – Endringer i jordbrukets kulturlandskap i Østfold', Oslo: NJOS, Akershus og Vestfold (2005), Tema arealstruktur (NIJOS rapport 12/05).
Fosso, E.J. (1998), 'Industriungdommen – blir de, reiser de og kommer de tilbake?' Upublisert innlegg på flytteseminar 19. januar 1998, Oslo.
Grønbech, D. (2000), 'Kvinneliv ved en skillevei', in M. Husmo and J.P. Johnsen (eds), *Fra bygd og fjord til kafébord?* Trondheim: Tapir Akademisk forlag, pp. 21–38.
Grønlund, A. (1989), 'Jordressursdata på nasjonalt nivå', Sluttrapport fra et prosjekt utført på oppdrag for miljøvern departementet, Rapport NIJOS.
Hänninen-Valjakka, Kirsi (1998), 'Etelä-Savon perinnemaisemat', Alueelliset Ympäristöjulkaisut 87, Etelä-Savon Ympäristökeskus, Mikkeli.
Information Centre of the Ministry of Agriculture and Forestry (2002), 'Agricultural Census 2000', Agriculture, forestry and fishery 2002:51, Helsinki.
Mäntyharju municipality (2005), Available at: http://www.mantyharju.fi.
Ministry of Agriculture and Forestry (2001), 'Strategy for Finnish Agriculture', Final Report of the Steering Group, Helsinki.
Ministry of Agriculture and Forestry (2003), 'Strategy for renewable natural resources in Finland', Available: http://www.mmm.fi/english/landwater/natural_resources/Luonnonvara_Englanti.pdf.
Mittenzwei K., M. Loureiro, W. Dramstad, W. Fjellstad, O. Flaten, A.K. Gjertsen and S.S. Prestegard (2004), 'A cluster analysis of Norwegian municipalities with respect to agriculture's multifunctionality', NILF-notat 2004-22.
MTT (2003), 'Finnish agriculture and rural industries 2003', Helsinki: MTT Agrifood Research Finland, Economic Research Publications 103a.
MTT (2004), 'Finnish agriculture and rural industries 2004', Helsinki: MTT Agrifood Research Finland, Economic Research Publications 104a.
MTT (2005), 'Finnish agriculture and rural industries 2005: ten years in the European Union', Helsinki: MTT Agrifood Research Finland, Economic Research Publications 105a.
National Board of Agriculture (1970), 'Agriculture. The findings by municipality', Official statistics of Finland III:67, Helsinki.
National Board of Agriculture (1983), 'Farm Register. Official statistics of Finland', XLIII:1, Helsinki.
National Board of Agriculture (1992), 'Agricultural Census 1990. The findings by municipality', Agriculture and Forestry 1992:6, Helsinki.
Nertsen, N.K., O. Puschmann, J. Hofsten, A. Elgersma, G. Stokstad and R. Gudem (1999), 'The importance of Norwegian agriculture for the cultural landscape: a subproject under the Ministry of Agriculture's evaluation programme on multifunctional agriculture', NILF-notat 1999:11.
NOU, Norges offentlige utredninger (2004), Livskraftige distrikter og regioner. Rammer for en helhetlig og geografisk tilpasset politikk. Statens forvaltningstjeneste, Informasjonsforvaltning, NOU 2004:19.
Norwegian Statistics (2002), *Statistical Yearbook*, www.ssb.no/aarbok/2002.
Orderud, G.I. (1998), 'Flytting mønstre og årsaker. En kunnskapsoversikt', NIBR, Prosjektrapport 1998:6.

Peltola, A. (2000), 'Viljelijäperheiden monitoimisuus suomalaisilla maatiloilla', Maatalouden taloudellinen tutkimuslaitos, Julkaisuja 96, Helsinki.

Rikkinen, K. (1992), 'A geography of Finland', Lahti: University of Helsinki, Lahti Research and Training Centre. ISBN 951-45-5887-1, SVT Agriculture, forestry and fishery 2003: Vol. 45, Helsinki.

Rogstad, B. (ed.) (2003), 'Norwegian Agriculture. Status and Trends 2003', Oslo: NILF, NILF publication.

Romstad, E. (2004), 'Methodologies for agri-environmental policy design', in F. Brouwer (ed.), *Sustaining Agriculture and the Rural Environment; Governance, Policy and Multifunctionality*, Cheltenham, UK and Northampton, MA, USA: Edward Elgar Publishing, pp. 56–77.

Statistics Finland (1960), *Statistical Yearbook of Finland 1960*, Helsinki.

Statistics Finland (1970), *Statistical Yearbook of Finland 1970*, Helsinki.

Statistics of Norway (2003a), Oslo. Naturressurser og miljø, http://www.ssb.no/emner/01/sa_nrm/arkiv/nrm2003/.

Statistics of Norway (2003b), Oslo. http://www.ssb.no/kommuner.

Tynset municipality (2003), http://www.tynset.kommune.no.

Vihinen, H., M.-L. Tapio-Biström and O. Voutilainen (2005), 'Rural marginalisation and multifunctional land use in Finland', Agrifood Research Working papers 103 (2005).

6. Marginalisation of rural economies in the Czech Republic and Hungary

Alajos Fehér, Josef Fanta, Gábor Szabó and František Zemek

The Czech Republic and Hungary, as former socialist countries, have faced radical changes in their political and economic systems since 1989. The change from the command economy to the market economy made the restitution of private land ownership possible. In 2004, the Czech Republic and Hungary became members of the European Union (EU) and adopted its agricultural policy. This introduced a period of instability resulting from the new economic conditions, the need to react to the Common Agricultural Policy (CAP) of the EU, competition in EU markets, and so on. Within a short time and at high speed, national agricultural policies and their new structures had to be developed to enable the agricultural sector to adapt to changing economic conditions. During this process, the agricultural sector faced a major transition in organisation, business, production and marketing, accompanied by marginalisation of rural areas and structural changes in employment. The process of changes in rural areas is still by no means complete. On the contrary, the countries involved have to reckon with long-term structural changes to bring the transition process to a satisfactory end.

The conditions in the Czech Republic and Hungary are different. They are deeply rooted in the history of these countries and in the different patterns of land use that evolved from specific natural, economic and social conditions.

The transition changed the socio-economic conditions in rural areas. People living in rural areas faced instability, due to a decrease in employment and uncertainty about household income. Young people also moved to cities in search of work. Shortage of capital, ageing of the rural population and the depopulation of rural areas have contributed to the marginalisation of land, and a decline in the viability of some rural communities.

The marginalisation of agriculture and rural areas is a very complicated and complex process. Although there are general political and economic causes, it also has very specific local and regional features. These depend on the local and regional environmental conditions such as geomorphology, climate, hydrology and soil quality, local demographic characteristics and population structure, and/or regional economic conditions.

This chapter provides an analysis of the transition process that took place in two countries and its consequences for the marginalisation of agriculture and rural areas. The features of marginalisation are analysed, and their consequences for the rural economy are indicated. In the framework of the investigation an important role was assigned to the multifunctionality of land use and agriculture. Based on this analysis, some conclusions are formulated on how to cope with the marginalisation of agriculture and how to increase the viability of the Czech and Hungarian rural economies under transitional conditions.

THE PROCESS OF MARGINALISATION

The general understanding of 'marginalisation' seems to be identical in both countries, but the specific interpretation tends to differ. It seems that in the Czech Republic the term was first used by people studying rural and regional development, especially related to land abandonment (Lipský, 1992; Bartoš et al., 1994; Mejstřík et al., 1995; Hanousková, 1998a). The cross-border areas were made especially inaccessible by the previous regime. The landscape ecological consequences of long-term abandonment received great attention (Hanouskova, 1998b). In these studies, marginalisation was understood as a consequence of long-term processes of changing landscape components and their economic, social, cultural and landscape roles. Evaluating the ecological aspects of the matter, Hanousková et al. (1999) concluded that marginal areas were the product of changing management systems. A monofunctional practice did not support sustainability and could result in marginalisation when economic conditions changed.

Negotiations at the end of the 1990s on the accession of the country to the EU, and the inevitable consequences for agricultural policy of the expected transformations, naturally stimulated research devoted to social and economic aspects of marginalisation (Koutný and Vaishar, 1997; Vaishar, 1999).

Szűcs et al. (1999) used statistical methods to identify areas with unfavourable conditions in Hungary. Fehér (2000) connected marginalisation with the term 'areas in double trouble', and he identified the factors which are causes of the permanently disadvantageous situation of the areas in question. The author explored in detail connections between unfavourable agricultural endowments and the cumulative disadvantages of regions. Baranyi (2001) reported numerous signs indicating that existing regional differences were likely to be aggravated, or at least remain unchanged, and that ever larger areas of the Great Hungarian Plain were in danger of becoming peripheries including border areas, micro-regions and individual settlements. The author noted that districts near the Hungarian–Ukrainian border, and in some cases near the Hungarian–Romanian border, were probably the most underdeveloped border regions from the social, economic and infrastructure point of view.

The economic and political changes of the 1990s not only opened the way for the marginalisation processes that had been controlled or suppressed earlier, but also caused the initiation of new regional processes. Serious economic and social tensions arose, combined with crises in the regional economies of the economically underdeveloped and peripheral areas. These developments obviously led to the acceleration of marginalisation (Table 6.1).

The process of marginalisation is most clear at the local level in the study areas[1] (Table 6.1). Between 1990 and 2000 the decline in the population was the highest in these areas compared to the NUTS 2 regions surrounding them and to the country levels. As regards the ratio of women in the total population, the value in Hungary as a whole was higher than in the Czech Republic, while in the Berettyóújfalu micro-region it was lower and had dropped to below 50 per cent by 2000. The absolute number of economically inactive individuals and their proportion within the working-age population increased substantially between 1990 and 2000 at all levels of investigation, but increased most in the micro-regions. The taxable income was low in study areas.

The major driving force of these processes is a radical increase in regional differences within the social and economic dimensions, while the measures taken to handle the situation were inadequate. These differences were primarily related to the diverse natural conditions and inadequate utilisation of land, the status of human resources, the state of development of social capital and of other resources, and the level of competitiveness of the regional economies. Also, they are caused by the peripheral nature of most rural areas, the protracted transformational and structural crisis caused by the change in the socio-economic regime, and the poor adaptability of people to the changes and other external and internal challenges.

The situation of the study areas demonstrated in Table 6.1 is not only characteristic for those two areas, but also for others. This is illustrated in Figure 6.1 for Hungary. The most advanced marginalisation processes occur in border areas with the exception of the Austrian border and the North Transdanubian section of the Slovakian border; in rural areas of the Northern Great Plain and Northern Hungary (which belong to the ten poorest NUTS 2 regions in the EU-25); and in underdeveloped rural areas of Southern Transdanubia.

In Figure 6.1 almost all of the marginalised micro-regions are located in rural areas and the share of utilised agricultural area in total territory and employment in agriculture are above the national average (without Budapest). In 2000 these figures exceeded the national average by 7.3 and 7.6 per cent respectively. The per capita income in these regions is only 61 per cent of the

Table 6.1 Changes in the Czech Republic and Hungary between 1990 and 2000 at country, regional and local level

Indicators for marginalisation	Czech Republic						Hungary					
	Study area		NUTS 2 region surrounding study area		Country		Study area		NUTS 2 region surrounding study area		Country	
	1990	2000	1990	2000	1990	2000	1990	2000	1990	2000	1990	2000
Population density (persons per km²)	44.6	43.1	66.9	67.1	130.6	129.0	58.2	56.3	87.3	85.8	111.5	108.0
Share of women in total population (%)	51.8	51.6	51.1	50.9	51.4	51.3	50.4	49.4	n.a.	51.4	n.a.	52.3
Share of economically inactive individuals in total population of working age (%)	14.7	23.4	14.1	21.1	13.9	21.4	27.7	47.1	27.7	42.9	22.1	35.2
Annual taxable income per permanent resident as % of the national average	n.a.	94.9	n.a.	89.9	n.a.	100.0	n.a.	31.8	n.a.	74.3	n.a.	100.0

Sources: The Czech Statistics Year Books 1991 and 2001; KSH (Hungarian Central Statistical Office) (2001a, 2001b, 2003a, 2003b, 2004a).

national average (without Budapest) and the density of enterprises only 59 per cent of the national average. These areas have a high proportion of economic inactivity and unemployment, with absolute values 3.5 and 4.9 per cent higher, respectively, than the national average (without Budapest). It should be noted that in Hungary the unemployment rate in itself does not demonstrate the actual process of marginalisation. The trend in the ratio of the economically inactive population was found to be much more relevant. The high economic inactivity, the very low taxable income per capita, poverty, and the high importance of the informal economy, all play an important role in the marginalised areas in Hungary. This indicates that the usual economic means and indicators are not always sufficient for an exploration of the processes and degree of marginalisation.

The damaging effects and threats of marginalisation have consequences for the viability of these areas in both countries. At present only a few micro-regions and settlements are affected by a population exodus. Nevertheless, due to ageing, population decline, poverty and deteriorating health, fundamental human resources are under serious threat, particularly in the most disadvantaged, underdeveloped regions.

The weakness of social capital, which is generally characteristic of areas struggling with marginalisation and the lack of cooperation, are particularly serious problems. A number of studies (Szabó, 2002; Szabó and Bárdos, 2005) have pointed out the lack of organisations and cooperatives for farmers, which could have played a significant role in the integration process through building up countervailing power.

Disadvantages arising from the basic poverty of natural resources are further exacerbated in these areas by the neglect of ecological and biological resources. Due to the permanently weak competitiveness of the economy, rural areas are also at a disadvantage as regards the acquisition and utilisation of knowledge, which substantially narrows their chances of achieving long-term development. The poor development of human and social capital as well as of regional economies are crucial factors hampering agriculture from undertaking functions other than commodity production (for example environmental protection and nature conservation, tourism, landscape protection and so on).

NATURAL CONDITIONS AND MARGINALISATION

The Czech Republic and Hungary differ substantially in their natural conditions. The landscape of the Czech Republic is very varied from a geological and geomorphologic point of view, consisting of lowlands, uplands and high mountains. Altitudes vary from 120 to 1600 metres above sea level. Climate is transitional, from sub-atlantic to sub-continental. Soils are very variable, ranging from rich chernozems, rendzinas and loess soils to very poor and gravel rich acid podzols. On the contrary, Hungarian landscapes are mostly large-scale flatlands, with mountains and hilly areas in Transdanubia and Northern Hungary. Soils vary greatly – from rich chernozems to sandy and saline soils and solonchaks and humic gley soils in lowlands or different types of forest soils in the hilly areas.

1 = Central Hungary; 2 = Central Transdanubia; 3 = Western Transdanubia; 4 = Southern Transdanubia; 5 = Northern Hungary; 6 = Northern Great Plain; 7 = Southern Great Plain. The grey areas are the most marginalised micro-regions. The spatial units were delimited by Governmental Decree No. 244/3003.

Source: Fehér (2005).

Figure 6.1   The most marginalised micro-regions in Hungary at the turn of the millennium

The Hungarian climate is largely continental. The differences in natural conditions are at the root of the very different forms and methods of agricultural land use in the past and in the historical development of rural areas in the two countries. The most important indices characteristic of land use and how it has changed between 1990 and 2000 are summarised in Table 6.2.

Table 6.2   Land use features in the Czech Republic and Hungary

Indicator	Czech Republic[a]		Hungary[b]	
	1990	2000	1990	2000
Share of uncultivated arable land (%)	n.a.	0.5	n.a.	6.7
Range of the share of uncultivated arable land at NUTS 2 level (%)	n.a.	0.3–1.4	n.a.	3.1–10.8
Stocking density (animal units per hectare of agricultural land)	1.2	0.6	0.4	0.3
Share of ruminants in total livestock population (%)	74.9	63.7	55.9	48.9
Share of protected area in total territory of the country (%) (excluding NATURA 2000 regions)	n.a.	15.8	n.a.	9.9
Share of LFA in total agricultural area (%)	n.a.	50.5	n.a.	15.1

Note: n.a. = not available.

Sources: [a] The Czech Statistics Year Books 1991 and 2001; [b] KSH (Hungarian Central Statistical Office) (2001a, 2001b, 2003a, 2003b, 2004a).

At the turn of the millennium, the share of uncultivated arable land in Hungary was very high. In contrast, this percentage was low in the Czech Republic. Special attention should be paid to the fact that the extreme values are high in Hungary even at the NUTS 2 level, reaching as much as 20 per cent in some micro-regions. In 1990 the stocking rate in the Czech Republic was three times that in Hungary. Livestock number declined in both countries during the 1990s, but the rate of decline was slower in Hungary. In both countries the grasslands are poorly exploited, as indicated by the decrease in the absolute number of ruminants and in their ratio within the total livestock number.

Considering the differences between the two countries with respect to relief and other natural conditions, and as regards the economic structure, the factors and processes behind the various indicators will be discussed separately.

Factors playing decisive roles in the marginalisation of agricultural lands in the Czech Republic are the very variable relief and productivity of soils. Due to the restrictions of relief, climate and soil, about half of the land used agriculturally is classified as Less Favoured Areas (LFAs) (Figure 6.2).

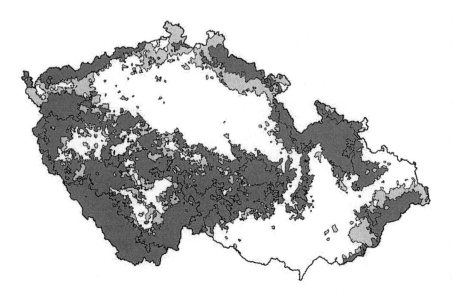

Note: Light grey areas are areas with environmental restrictions.

Figure 6.2   Less Favoured Areas in the Czech Republic

In the past few decades, Czech landscapes suffered greatly from explicitly environmentally unsound land use management. The ecological losses included a pronounced decrease in biodiversity, loss of biotopes and habitats, and decreased water storage capacity of the land (Kender, 2004). The agricultural sector has experienced considerable reduction in land use in the first ten years (1990–2000) of the transition period (and further changes are expected in the future), while landscape protection and nature conservation have increased in the Czech rural landscape in these years. This is a positive sign; the protected areas have a stabilising effect on the rapidly changing situation and can reduce the harmful effects of marginalisation. The Czech Republic already has a good network of both large-scale protected areas (four national parks, 25 landscape protected areas) and small-scale nature reserves spread over the whole country, protecting and conserving specific natural biotopes and ecosystems, valuable plant, animal or insect species and various abiotic phenomena. In total, some 16 per cent of the national area falls under one or another form of nature or landscape protection (Forestry Section, 2003). Restoration of those areas will not only have a very positive effect on landscape, but will also be an important source of employment for the rural population. Comprehensive planning will be essential to establish robust ecological networks, and to restore biodiversity and water retention capacity of the landscapes in question. The Czech protected areas form an essential part of the European Ecological Network. Many of them have been included in the Natura 2000 programme and network (AOPK, 2005).

In Hungary unfavourable soil and hydrological conditions, and the hilly character of the landscape, play a decisive role in the marginalisation of agricultural lands. The acidic brown forest soils with shallow topsoil particularly in mountainous and hilly areas, and the blown sands, saline solonchaks and solonchak-solonetzes in the Hungarian Great Plain, increase the vulnerability to marginalisation due to the low productivity of these soils under current socio-economic conditions.

Ameliorative liming or treatment with gypsum to improve the soil was applied in these areas during the 1970s and 1980s, but has greatly declined since the 1990s, more so in areas threatened by marginalisation than in other regions. Another impact of soil on vulnerability to marginalisation is caused by drought on highly permeable and poorly water-retentive soils like sandy soils low in organic matter, and on wet soils. These unfavourable hydrological features were partly improved by soil loosening and drainage during the 1970s and early 1980s, but these installations were amortised after the change of regime. Soils with such unfavourable conditions for arable farming are often used as meadows. However, due to the strong decline in stocking density these lands are not used in an optimal way and make farming on these lands not profitable.

In hilly areas the combination of slope, the low level of organic matter resources and water erosion accelerates marginalisation of agricultural lands due to low productive soils, while wind erosion has a similar effect on the blown sandy soils of the lowlands. Water and wind erosion is closely connected with inappropriate land cultivation: that is, machine hoeing, ploughing in the direction of the slope, too-frequent ploughing, bare soil in early springtime, and so on. Such practices became more frequent in the 1990s. According to foreign and Hungarian studies the sloping nature of these areas (with gradients exceeding 5 per cent) is one of the most important factors in the marginalisation of traditional agriculture in Central and Eastern European countries.

In Hungary the radical reduction in livestock and the chaotic state of land ownership and land use resulted in considerable areas of abandoned grasslands and long-term fallow. These problems were the gravest in border areas and inner peripheries. The process of abandonment of land use is accompanied by serious economic difficulties and by population migration, which also jeopardise these ecologically valuable areas. The enumerated factors characteristically aggravate the marginalisation of agricultural land use in LFAs (Figure 6.3). Disadvantageous endowments multiply each other's impacts and may lead to cumulative drawbacks.

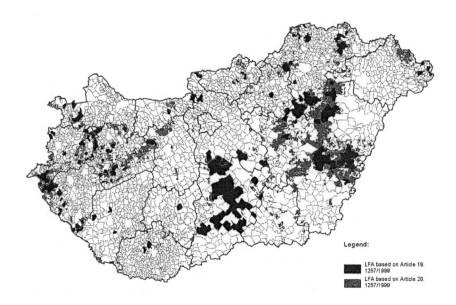

Legend:

LFA based on Article 19. 1257/1999
LFA based on Article 20. 1257/1999

Figure 6.3   Less Favoured Areas in Hungary

Based on the recommendation of the European Commission regarding selection criteria for Article 20 LFAs, Hungarian Article 20 areas were designated using four specific handicaps (agronomic limitation factors): severe acidity, severe salinity, extreme water management (inundation, water-logging) and extreme physical characteristics of soils. As the number of LFAs determined on the basis of the various criteria for natural disadvantages, separately or cumulatively, would exceed 10 per cent of the country's territory (the limit laid down by EU Regulation 1257/1999), areas where at least two of the four criteria were satisfied simultaneously were chosen as eligible under Article 20 (Ministry of Agriculture and Rural Development, 2004).

The criteria for the determination and delimitation of Hungarian LFAs appear to have been applied too strictly and in a contradictory manner. Many hilly regions have been omitted from the list, as have the majority of marginalised areas, including border regions. The real extent of LFAs is far greater than that indicated in Table 6.2 and Figure 6.3, and can be attributed both to unfavourable natural endowments and to socio-economic conditions.

The delimitation of Natura 2000 areas[2] could partly counterbalance these problems, since 14.7 per cent of the total territory of the country is included in this category.

Since 1980 the cultivated agricultural area has declined in favour of forests and uncultivated land in Hungary. The status of agricultural land (especially for grassland, vineyards and orchards) is much worse than it was in the 1980s. For instance the Ministry of Agriculture and Rural Development found that only 48.1 per cent of Hungary's total grassland satisfies the criteria for good agricultural practices. In recent decades less land has been removed from agricultural use in marginalised micro-regions than in more developed regions of Hungary. On the other hand, a considerable amount of long-term fallow has replaced ploughed areas, while large areas of grassland are unused and neglected in marginalised areas. At the turn of the millennium 6.5 per cent of the arable land in Hungary was long-term fallow, half of this being found in the two macro-regions most seriously affected by marginalisation (Northern Hungary and the Northern Great Plain), which contained 33 per cent of the total arable land in Hungary at the time in question. At the same time the rate of afforestation slowed in marginal areas, with a radical drop in the area planted and deterioration in the quality of the forest stand.

ECONOMIC IMPACTS OF MARGINALISATION

Compared to the more stabilised situation in agriculture in the EU-15 countries (where political development was continuous in the course of the twentieth century, though changes were wrought by the need to adapt to the world markets and economy), rural economies in the Central and Eastern European (CEE) countries face a very complicated situation. On the one hand, rural economies need to be re-established, based on democratic rules and private land ownership, and stabilised properly to fulfil their tasks as part of national economies. On the other hand, as part of the European agricultural economy, the national agricultural sectors must be involved in and adapted to the CAP and to its macroeconomic, environmental, socio-economic and socio-cultural principles. In this way marginalisation is mainly a socio-economic issue in the Czech Republic and in Hungary (Szabó and Fehér, 2004).

The traditional role of agriculture and forestry in employment has changed substantially, and as already mentioned, the contribution of agriculture to employment has dropped considerably. The Czech agricultural sector employs 4.2 per cent of the active population, and more than half of them are aged 45 years or more. In the transition period the level of employment decreased dramatically by 73 per cent. In Hungary large-scale farms provided a considerable number of jobs for the rural population until 1990, with 15.5 per cent of the total number of employed working in agriculture and forestry.

By 2000 the ratio of people employed in these sectors had dropped to 5.6 per cent, 42 per cent of whom were entrepreneurs and farmers and 58 per cent paid employees. The majority of agricultural workers made redundant between 1990 and 1994 became unemployed, and only around a tenth set up independent businesses. Only a small proportion of the unemployed were later able to make an independent living, while most are dependent on casual work, social security or income from the black and grey economies.

The territorial distribution of people affected in some way by agriculture differs greatly in different parts of the country, with a higher percentage in macro-regions with a large proportion of marginalised micro-regions. In rural settlements in Northern Hungary and the Northern Great Plain, for example, small and medium-sized farms (10–300 hectares) employed 12 per cent of the total adult population (over the age of 15), calculated in annual work units. The fate of these people is greatly endangered by the increasing rate of farm concentration, the main driving forces for which are efficiency requirements, technological development and the requirements of economics of scale. As a consequence of farm concentration it is likely that Hungarian agriculture will make more people redundant after EU accession than in previous years. When smallholders stop farming even part-time and sell or rent out their land, they and their family members enter the job market in rural economies (Fehér, 2005). As yet, however, there are no jobs available in other sectors of the rural economy in marginalised micro-regions to absorb these new job seekers.

In both countries two mainly market-driven trends can be perceived at this stage of development:

• Areas favourable for agriculture. An ever-increasing intensity and scale of land use is leading to better economic results, but simultaneously resulting in less employment for the rural population, and sometimes exceeding the carrying capacity of the landscape (for example excessive erosion). Agriculture is not only the most important land use category, but it is also one of the major economic sectors in rural economies. As a whole, an increase in the volume of traditional agricultural outputs (food and raw products) and employment can be expected due to a rise in yields, to varying extents in different sectors. However, this increase in production is likely to be more rapid in better-endowed regions. This will have negative consequences for farmers in marginalised areas, as regards loss of markets, the distribution of production quotas and subsidies linked to basic production.

• Areas with less-favourable conditions. Such regions face a risk of extensiveness of agriculture and land abandonment, with potentially beneficial impacts on the environment, but accompanied by the marginalisation of these areas.

After the political changes, a very complicated, bipolar landownership and business form structure has been observed in both countries. Recent land-ownership, as well as farming on hired land, are experienced as complicating factors in the consolidation of the agricultural sector. Also, the extended farm lease system in its present form does not create the best conditions for proper soil and land management. Another complicating factor is the unstable structure of business forms in Czech agriculture. Large companies reveal a greater capital security, leading to increased investments, modernisation, innovation and competitive power. On the contrary, the capital weakness of small companies and private farmers is the essential factor that determines whether the business survives and continues. Two economic trends are observed:

- Decline in number of private farmers; in most cases, large cooperative and agricultural companies are taking over their land.
- Slow decrease in the number of cooperatives, which are often changed into companies.

Both trends increase the scale and intensity of land use. These trends may increase financial advantages for the large companies, in both favourable and less favourable areas. They also increase the disadvantages for the landscape and for rural areas (ecological setbacks in the form of decreased soil quality, increased erosion and/or soil compaction, decreased landscape biodiversity, and less employment for rural populations). To counter these trends, greater support for the small and medium-sized private farmers is expected in the coming years (MZe, 2005).

In Hungary there are numerous obstacles to the rationalisation of land use:

- The underdevelopment of the economy of the micro-region (scarce human resources, predominance of agricultural production, unfavourable conditions for agriculture, weaknesses in various sectors of the regional economy, deficiencies in local economies, the low level of agricultural farm diversification).
- Repercussions of deficiencies in national policies aimed at stimulating the development of regional economies before accession to the EU. The strongly horizontal and sectoral features of national documents aimed at the implementation of the CAP (the Agricultural and Rural Development Programme of the National Development Plan, National Rural Development Plan) instead of regional dimensions and the insufficiency of available financial resources. In Hungary there is still a high level of centralisation; in some cases, due to repeated reorganisation, the administration of the micro-regions has become extremely uncertain, while the macro-regions do not have enough scope of action or sufficient

financial resources. The local governments of underdeveloped areas are poor; since they depend financially on the central government, they cannot become completely independent. The level of strategic planning and willingness to cooperate is low.

All these factors have contributed to the fact that the development documents have a sectoral approach which is reflected in their aims and structures. Even in the Regional Operative Programme the same features can be discovered, rather than the objective of tackling the actual problems of the given regions.

The problems differ in both countries, depending among others on the physical-geographical and environmental conditions, the general economic circumstances and the local demographic situation. In some areas, agriculture itself is responsible for some of the driving forces of marginalisation (for example a low level of entrepreneurship and innovation in the rural population). On the other hand, agriculture as a land use form and economic sector also suffers from the social and economic marginalisation of certain regions. The abandonment of agricultural land and the introduction of various forms of extensive land use (for example grasslands instead of arable land) may be beneficial for ecosystem processes and the biodiversity of landscape, and could improve the rural resources. However, the negative consequences of economic marginalisation can only be reduced over a longer term. On the other hand, extensive land use is often accompanied by less employment which has adverse effects on the socio-economic conditions of rural population and continues to feed the economic forces driving marginalisation. Traditional agriculture is unable to resolve this contradiction.

MULTIFUNCTIONALITY TO COPE WITH MARGINALISATION IN RURAL AREAS

The present inability of conventional agriculture to make a satisfactory contribution to the stabilisation of rural economies in marginalised areas can be attributed to the fact that the large-scale farms previously operating in these micro-regions have gone bankrupt and the new farms are unable to take over their role in the changed economic and job market environments. Although many people have suffered from the changes, there has been an unexpected benefit for the natural environment (Kerekes, 2003). Since small and medium-sized farms in LFAs are farmed less intensively, with fewer inputs, there has been a reduction in the environmental pollution caused by agriculture. This effect has been further strengthened by agricultural and environment management measures. It can be demonstrated, for instance, that projects sponsored by the Hungarian National Agricultural and Environment Management Programme were mostly located in zones with non-intensive agricultural production and played an above-average role in environmentally sensitive areas (Katonáné Kovács, 2005).

On areas with poor agricultural potential, particularly threatened by marginalisation, the environment management activities of farmers become far more important, and payment to them for environmental services could be a way of boosting the stability of rural economies.

Marginalisation and the demands for environmental and natural assets stimulate a change of paradigm in conventional agriculture. In the marginalised areas the sustainability of local resources and new economic activities have appeared as a real challenge. The multifunctionality of agriculture seems a feasible paradigm for coping with new challenges (for a detailed discussion on multifunctionality of agriculture, see Brouwer, 2004). Multifunctional land use was recommended as a possible remedy or tool to counteract marginalisation of rural areas in the Czech Republic and in Hungary. Contrary to the often-proclaimed limitations on land use in large protected areas, Těšitel et al. (2005) highlighted the opposite: a positive impact of a combination of tourism, recreation and nature protection on regional economic development in several large protected areas in the country. These results indicated one of the possible ways to counteract marginalisation under local and regional conditions.

Some key aspects of multifunctional agriculture in marginalised areas of the Czech Republic and Hungary will be discussed below.

Typical Czech LFAs (mountains and foothills) have relatively high potential for multifunctional agriculture and forestry, which has already been partly recognised by the local population and stakeholders, and partly put into practice. At the same time, it has become clear that a change in agricultural land use can only offer limited possibilities to improve both the situation of land users and the viability of the area.

Based on local traditions and the available natural and human sources, a further shift in agriculture from common crops and meat production to other possibilities, including small-scale farming, agritourism, ecological farming, and biomass for energy production seem to be most promising. Also, forestry, recreation, tourism and related services, small-scale industrial and handicraft activities, and landscape restoration and management can be attractive options.

Support for small-scale family farms is highly desirable, and is likely to have a positive influence on landscape structure. The situation of small farms is not easy because large agro-concerns dictate the buy-up prices of many products, and a market for higher-quality products from ecological farms and small-scale farms has not yet been developed. Lack of cooperation among farmers and appropriate marketing structures are weak points. Subsidies to small farmers oriented towards diversification of land use, as well as support (finance, know-how) to establish appropriate marketing structures, will certainly be beneficial for the region in coping with marginalisation.

Agritourism has recently become a leading activity because there are numerous clients from large towns. Many mountain and foothill areas in the Czech Republic are attractive to visitors, offering them areas of preserved nature, a beautiful landscape, rivers for canoeing, many famous historical monuments, and many unspoiled towns and villages. Unfortunately, this high potential is still not adequately exploited. A better advertising policy together with infrastructure development, such as connecting interesting and valuable sites by marked cycle or footpaths and establishing small guesthouses in rural areas, should enhance the economic role of agritourism. This has started in some parts of the country.

Under Czech conditions, growing biomass for energy production in LFAs can be an interesting alternative, when combined with extensive grazing and nature protection (grassland diversity). A small-scale approach and fermentation or gas production technology should be preferred rather than industrial methods. Two obstacles are restricting a broader use of this technology:

- The monopolistic position of the Czech energy producer (CEZ) on the energy market.
- The lack of investments in rural areas to build the necessary installations. A political decision will be needed to facilitate the matter. The Horizontal Rural Development Plan for the period 2007–13 involves some steps in this direction (MZe, 2005).

In the past, forestry was an important source of seasonal employment in mountain regions. These possibilities have been reduced considerably under contemporary conditions. The large-scale management approach of forestry companies does not support local employment opportunities. Nor is it in line with the recent trend towards forest sustainability and multifunctionality, achieved by the re-establishment of the natural state or continuous-cover forests. A change in this field will depend on political decisions, which will streamline the development of Czech forestry in accordance with European forestry policy and can bring new jobs.

Decreasing the scale of forestry companies can also be beneficial for the development of small-scale industrial and handicraft activities in the area, producing the required materials for households and garden maintenance. A further possibility would be the outplacement of workshops from urban to rural areas. A financial stimulus by regional governmental bodies, in the form of tax reductions, could be an attractive alternative.

The development of multifunctional agriculture in Hungary is still in its infancy. First, in 2002 the National Agricultural and Environmental Protection Programme was introduced, and the measures laid down in the National Rural Development Plan came into force in 2004, containing substantial incentives to farmers providing environmental services. In 2002 and 2003, 4 per cent of Hungary's agricultural area was involved in both programmes (Katonáné Kovács, 2005).

Hungary is not particularly well endowed for the utilisation of wind and solar energy, so the participation of agriculture in supplying energy is most conceivable in the form of biomass. It has been calculated that the energy-producing potential of agriculture and forestry is more than three times its own fossil fuel requirements. Although the Hungarian government has stated its intention of supporting an increase in the proportion of electrical energy and fuel produced from biomass, in reality only a small number of developments have emerged, all of purely local significance. In addition, the cultivation of known plant species suitable for energy production is likely to be more efficient on more fertile areas, so it is here that their wide-scale production is likely to start when payments become available.

On-farm processing of agricultural products is of limited importance. In 2003 only about 0.5 per cent of private farms were involved in the processing of meat, milk, vegetables or fruit. In the case of agricultural companies this percentage was somewhat higher (around 0.8 per cent) for meat, vegetables and fruit, but even lower (0.4 per cent) for milk (KSH, 2004a).

So far, activities such as village and farm tourism remain of limited importance. Even in 2000 the proportion of farms involved in this activity was very low, and between 2000 and 2003, the figures dropped to half for private farms and to two-thirds for agricultural companies (KSH, 2004b). Organic farming is also marginal. The proportion of authorised organic farms and farms under conversion was less than 0.5 per cent in 2000, but it is encouraging that a further 1.2 per cent of the farmers are considering introducing organic farming methods.

In the micro-regions that are affected most by marginalisation, agricultural and environmental management measures and organic farming are more widespread than in the rest of the country. In contrast, this is not observed for other types of multifunctional activities at the turn of the millennium.

Fehér and Biró (2005) distinguish the following measures for improving the multifunctionality of Hungarian agriculture:

- Tasks concerned with the supply of food, raw materials and energy.
- Improvements in employment and living standards through farm diversification.
- Rural landscape protection, the maintenance and improvement of biodiversity through the provision of nature conservation and other environmental services.
- Contributions to the stability of rural economies and more uniform territorial development.

Realisation of these tasks in marginal areas could be improved by the change in the economic and political environment since EU accession (an increase in agricultural payments; environmental regulations; agricultural, environmental and rural development programmes; regulations with regard to renewable energy resources; demand for healthy, safe foodstuffs).

The greater value paid to the landscape and biological resources on an international scale, and examples of their utilisation in EU-15, is important for increasing multifunctionality. Another point is to reduce the difficulties in placing traditional agricultural products and services on Hungary's usual markets, and the increasingly great risks involved in new markets.

Developing cooperation between border regions in Hungary was suggested too. The regional (rural) economies of marginalised border areas – for example, regions along the Hungarian–Romanian and Hungarian–Slovakian border – are in a special situation because of the potential impact of EU membership. Theoretically the membership of these countries could considerably counteract the marginalisation of the areas in question. Based on the traditions that existed before the Second World War and on natural spatial connections, as well as on the European regional trends, some experts are very optimistic about the possibility of economic cooperation and its dynamic and positive influence on some micro-regions (Baranyi, 2001). Without going into details, it seems unlikely that development will be rapid in the short time. This opinion is based on experiences in Slovakian–Hungarian cooperation in Northern Hungarian areas since EU accession and on the numerous economic, social and other barriers built during a half-century on both sides of the border. Although the artificially divided, peripheral areas have become part of new Euro-regions since the EU accession, there are no towns capable of stimulating regional innovation, so cooperation is difficult in these areas. The impact accession to the EU in counteracting marginalisation in these border regions can only be expected in the long term (Szabó and Fehér, 2005).

At the same time a number of conditions and factors have a profoundly inhibitory effect on the spread of multifunctional agriculture in marginal areas of Hungary:

- The adaptation of farmers to the changes was inadequate and their insistence on traditional farming is strong.
- The low social awareness of the use of alternative energy sources, the excessively bureaucratic regulations and the lack of the necessary technical conditions for the provision of such energy is slowing down the process. Further, the present status of rural resources and the lack of knowledge and experience of alternative utilisation, diversification, non-commodity outputs and alternative sources of income are obstacles. Also, the force of negative spatial processes (including marginalisation) and the underdevelopment and weakness of rural economies makes realisation of multifunctionality difficult.

- In addition, certain environmental measures (for example afforestation, formation of aqueous habitats) reduce the potential for food production for a longer period, making farmers and public decision-makers wary. A major share of the agricultural plots are too small and isolated, and in many cases owners and land users have conflicting interests. Also, an important fact is that human and social capital is poorly developed in marginal areas (for social capital see Chapter 11 of this volume).

CONCLUSIONS

In the countries analysed, political regimes artificially interfered in regional and economic processes until the late 1980s. As a result, the development of regional economies was retarded, while regional tensions and marginalisation processes were suppressed. Following that period, socio-economic processes accelerated to a great extent, partly in an attempt to catch up with the developed countries. This, however, led to regional tensions, crises and intense socio-economic marginalisation. Countries in transition were unable to adjust to the spatial processes of the market economy, lacking the political experience and tools required to influence these processes. The poorly developed regional and local economies, communities and rural population were completely unprepared for the complex challenges they suddenly had to face, largely because of the weakness of the civil society and social capital. Very few real results have been achieved in the easing of these tensions in recent decades, while more and more difficulties have arisen, all of which have made their mark on Czech and Hungarian research papers dealing with marginalisation. In earlier years, scientists only worked on specific parts of the problem, and it is only recently that complex research on marginalisation has begun.

The first experience following accession to the EU is that the Czech and Hungarian rural economies are still involved in a process of deep change – a transition from centrally planned to market-driven conditions. This process of change has had major impacts on rural economies and land use in rural areas, and will continue to do so in the coming years. The most important characteristics of the present stage in this transition are:

- social uncertainty, mistrust with regard to future development, capital weakness and lack of entrepreneurship in rural populations;
- lack of appropriate tools and experience of how to manage the change in regional and local governance bodies (such as regional and landscape planning, inter-sectoral cooperation and coordination); and
- uncertainty about the potential ecological impacts of the transition on rural landscapes and their carrying capacity.

In the Czech Republic, uplands and mountain areas along the borders with Germany, Poland and Slovakia are confronted with various forms and features of marginalisation. The major decline of employment in the agricultural sector during recent years was counteracted to some extent by employment in industry and services, largely in and around urban centres. Nevertheless, the development of new employment opportunities in rural areas remains a structural problem, which deserves concentrated attention.

In Hungary, regional differences in economic and social conditions have significantly increased since the change of the economic and political system in 1989. Marginalised areas can be found particularly in the Northern Hungarian, Northern Great Plain, Southern Great Plain and Southern Transdanubian regions. While marginalisation in the Northern Hungarian region is mainly due to the liquidation of heavy industry (metallurgy), and in the Southern Transdanubian region marginalisation is particularly a consequence of the liquidation of the mining industry, the permanent crisis of agriculture also plays an important role in the process. In the Northern and Southern Great Plain regions, the development of marginalisation is mainly due to the crisis in agriculture, accompanied by the significant decrease in food industry activities.

Since Hungary entered the EU, the situation of pig and poultry farms and of the dairy sector has significantly worsened, leading to a further decrease in livestock. These factors led to a further decline in the ability to maintain agricultural population, further strengthening the marginalisation of these areas. Rural development programmes have proved insufficient to countervail increasing employment problems here. In the majority of marginalised Hungarian areas agriculture is one of the main sources of livelihood, not necessarily in the form of full-time jobs, but as a supplementary source of income and food for rural households.

Under these circumstances, conventional agriculture is losing its ability to lead the way in maintaining the viability of rural areas. What is at stake is a decrease in the living standards of the rural population, or an exodus from rural areas in search for jobs elsewhere. The need to expand activities in the agricultural sector beyond the usual commodity production by diversifying farm activities, and the search for other income sources (for example offering accommodation for recreation; seeking off-farm jobs and so on) gained importance. The opportunities provided by European Multifunctional Agricultural Models in marginalised areas would significantly contribute to the termination of this contradiction and to the moderation of marginalisation.

The delimitation of marginalised areas at the national level and the framing of separate regional measures for these areas would appear to be indispensable. The harmonisation of LFAs, marginalised regions, Natura 2000 and other protected areas will require new approaches and methods. The coherence and synergic effects between various payments (LFA, agri-environment, diversification in agriculture and other special regional measures) is essential. Within the regional framework, the LEADER+ programmes, and especially projects aimed at developing the initiative of local communities, are of great importance in the new EU member states.

Significant changes can be expected during the budget period 2007–13. The regional approach to rural development plans and projects should gain a greater emphasis in both countries. To counteract the situation and its negative consequences effectively, the following steps are recommended:

- Development and diversification of regional economies. In micro-regions endangered by marginalisation the national development and rural development plans, prepared and divided according to regions, should emphasise the development of knowledge, the accentuated improvement of the service sector and the stimulation of economic relations with more developed areas. This solution could significantly increase both the value added per capita and the personal income in the micro-regions, as well as contributing to a reduction in economic inactivity. It could reduce the employment pressure on agriculture and could foster structural changes in this sector. The improved performance of the regional economy could raise living standards and make the settlements more attractive to intellectuals and educated people, thus contributing to other positive changes in the field of human resources.
- Diversification and multifunctional development of local agriculture. The abundance of ecological, biological and landscape resources together with the reform of the CAP in 2003 could stimulate these changes, such as for example an increase in the proportion of alternative crop production and stock-breeding, the encouragement of organic farming, the introduction of regionally specific products, and the production of biomass as a source of energy (biofuel sources of the second generation).
- Development of regionally specific services based on a thorough assessment of resources and possibilities. Due to regional differences, the possibilities will be very different and regionally specific, involving for example rural tourism, recreation on farms, local fishing and hunting, attending local celebrations (for example harvest festivals, annual festivals), and so on. Some public services are hard to measure, such as protection of the environment, or increasing and maintaining biodiversity. However, landscape conservation, restoration and management, the establishment of ecological networks and measures to increase the water storage capacity of the land must be seen as important sources of employment for the rural population.

Such measures could be a solid basis to cope with marginalisation of rural areas in the context of countries facing major reforms in the recent past.

NOTES

1. The micro-regions (NUTS 4 level) in question form part of the former district of Klatovy in the Czech Republic, and the Berettyóújfalu micro-region in Hungary. The location of the latter region is presented in Figure 6.1.
2. See Council Regulation (EC) No 1698/2005 of 20 September 2005 on support for rural development by the European Agricultural Fund for Rural Development (EAFRD). Article 50 of this regulation deals with the Natura 2000 agricultural areas designated pursuant to Directives 79/409/EEC and 92/43/EEC. The said areas shall be eligible for payments provided for in Article 36(a) (iii).

REFERENCES

AOPK (2005), *Natura 2000 Network in the Czech Republic. Map 1:500.000*, Praha: Agentura Ochrany Přírody a Krajiny.

Baranyi, B. (ed.) (2001), *A határmentiség kérdőjelei az Északkelet-Alföldön* (Border Issues in the North-Eastern Part of the Great Plain), Pécs: MTA Regionális Kutatások Központja.

Bartos, M., J. Tesitel and D. Kusova (1994), 'Large scale abandonment and its consequences', in A. Richling, E. Malinowska and J. Lechnio (eds), *Landscape Research and its Applications in Environmental Management*, Warsow: IALE, Warsaw University Press, pp. 213–20.

Brouwer, F. (ed.) (2004), *Sustaining Agriculture and the Rural Environment: Governance, Policy and Multifunctionality*, Cheltenham, UK and Northampton, MA, USA: Edward Elgar.

Datastar database (2001), Tarnamente, Tisza-tó ÖKO alapítvány, Poroszló.

Fehér, A. (2000), 'Halmozottan hátrányos térségeinkről' (Areas of Hungary with cumulative disadvantages), *Gazdálkodás*, 44, 68–79.

Fehér, A. (2005), *A vidékgazdaság és a mezőgazdaság* (Rural economy and agriculture), Budapest: Agroinform Kiadó, Vol. 186.

Fehér, A. and Sz. Biró (2005), *A multifunkciós mezőgazdaság megteremtésének esélyei és teendői* (The chances of creating multifunctional agriculture and the steps that should be taken), Kézirat, Budapest, MEH Integrációs és Fejlesztéspolitikai munkacsoport.

Forestry Section (2003), 'Report on the State of Forests and Forestry in the Czech Republic', Prague: Ministry of Agriculture, Report 114.

Hanousková, I. (1998a), 'Anthropoecological changes of Šumava Mts area, Czech Republic', in A. Richling, E. Malinowska and J. Lechnio (eds), *Landscape Transformation in Europe. Practical and theoretical aspects*, Warsaw: IALE, Warsaw University Press, pp. 158–63.

Hanousková, I. (1998b), 'Changes and reforms in agricultural landscape in Czech lands', in A. Richling, E. Malinowska and J. Lechnio (eds), *Landscape Transformation in Europe. Practical and theoretical aspects*, Warsaw: Warsaw University Press, pp. 281–6.

Hanousková, I., P.E. O'Sullivan and Z. Witkowski (1999), 'Perspective in sustainable use of marginal areas, land abandonment and restoration', in A. Farina (ed.), *Perspectives in Ecology*, Leiden: Blackhuys Publishers, pp. 295–308.

Katonáné Kovács, J. (2005), 'Az agrár-környezetvédelem és a vidékfejlesztés összefüggései az Európai Unióhoz történő csatlakozás tükrében' (Correlations between agricultural and environment protection and rural development in the light of EU accession), PhD thesis, Debrecen: Debreceni Egyetem Interdiszciplináris Társadalom- és Agrártudományok Doktori Iskola, Report 116.

Kender, J. (ed.) (2004), *Water in Landscape,* Prague: Consult Publishers.

Kerekes, S. (2003), 'A magyar gazdaság környezeti teljesítménye az átmenet korában' (Environmental performance of the Hungarian economy during the transitional period), DSc thesis, Budapest, Magyar Tudományos Akadémia.

Koutný, R. and A. Vaishar (1997), 'Transformation in marginal regions: the example of Middle Dyje Region', *Acta Universitatis Carolinae-Geographica*, Praha: Univerzita Karlova, pp. 357–67.

KSH (Hungarian Central Statistical Office) (2001a), 'Általános Mezőgazdasági Összeírás 2000' (General Agricultural Census 2000), Budapest.

KSH (2001b), 'Általános Mezőgazdasági Összeírás: Földhasználat Magyarországon a 2000. évben' (General Agricultural Census: land use in Hungary in 2000), Budapest.

KSH (2003a), *Magyarország régiói* (Regions of Hungary), Budapest.

KSH (2003b), *Területi Statisztikai Évkönyv, 2002* (Regional Statistical Yearbook, 2002) Budapest.

KSH (2004a), *Mezőgazdasági termelés, 2003* (Agricultural production, 2003), Budapest.

KSH (2004b), *Magyarország mezőgazdasága, 2003* (Hungarian agriculture, 2003), Budapest.

Lipský, Z. (1992), 'Long-term analysis of landscape development: Application in recovery or regional stability', Thesis, Institute of Applied Ecology, Kostelec nad Černymi lesy.

Mejstřík, V., M. Bartoš, I. Hanousková and J. Těšitel (1995), 'Abandoned landscapes in the Czech Republic. Šumava Mts. Case', in J.F.H. Schoute et al. (eds), *Scenario Studies for Rural Development*, Dordrecht: Kluwer Academic Publishers, pp. 39–45.

Ministry of Agriculture and Rural Development (2004), *National Rural Development Plan, Final version*, Budapest, 19 July, pp. 128–9.

MZe (2003), 'Horizontal Rural Development Plan for the Czech Republic, 2004–2006', Working document, Ministry of Agriculture ČR.

MZe (2005) 'Rural Development Program ČR', Working document, Ministry of Agriculture ČR, Prague (in Czech).

Szabó, G.G. (2002), 'New institutional economics and agricultural co-operatives: a Hungarian case study', in Simeon Karafolas, Roger Spear and Yohanan Stryjan (eds), *Local Society and Global Economy: The Role of Co-operatives*, ICA International Research Conference, Naoussa: Hellin, pp. 357–78.

Szabó, G.G. and K. Bárdos (2005), 'Vertical coordination by contracts in agribusiness: an empirical research in the Hungarian dairy sector', Budapest: MTA KTI, Discussion Papers series (DP-2005/15).

Szabó, G. and A. Fehér (2004): 'Marginalisation and multifunctional land use in Hungary', *Acta Agraria Debreceniensis*, Debreceni Egyetem ATC, 15, 50–62.

Szabó, G. and A. Fehér (2005), 'Marginalisation of agriculture in a Hungarian border micro-region', *Studies in Agricultural Economics*, 103, 53–71.

Szűcs, I., K. Daubner, M. Galó, M. Goda, A. Laczkó, A. Fehér, R. Magda, M. Spitálszky and A. Tenk (1999), 'A halmozottan hátrányos térségek.gazdasági társadalmi lemaradása' (Social-economic backwardness of cumulative disadvantageous regions), in F. Kovács, I. Dimény, and I. Szűcs (eds), *A mezőgazdaság szerepe a halmozottan hátrányos helyzetű térségek fejlesztésében*, Budapest: Magyar Tudományos Akadémia, pp. 57–85.

Těšitel, J., D. Kušová, K. Matějka and M. Bartoš (2005), 'Protected landscape areas and regional development (the case of the Czech Republic)', *Rural Areas and Development, Vol. 3, Rural Development Capacity in Carpathian Europe*, Warsaw: Institute of Agricultural and Food Economics and Institute of Geography and Spatial Organisation PAN, pp. 23–36.

Vaishar, A. (1999), 'Marginal regions in Moravia: transformation of the social and economic system and its consequences', *New Prosperity for Rural Regions*, Geographica Slovenica, 31, Institute of Geography, Ljubljana, pp. 102–16.

7. Rural amenities in mountain areas

Thomas Dax and Georg Wiesinger

Mountains cover a significant portion of the land area of many countries in the world, and their resources are playing a crucial role in sustainable development. Due to their remoteness great parts of them are threatened by marginalisation tendencies and the specific challenges of development in mountain areas are rarely reflected in national policies. Relevant issues include, in particular, situations of 'difficult access, economic and political marginality, out-migration, environmental sensitivity, diversity of livelihoods, and cultural diversity' (Mountain Agenda, 2002: 4). Despite these common challenges, activities in mountain areas and respective policies have to apply strategies and instruments which address the region-specific contexts and the great diversity within mountains.

Mountain areas are perceived as some of those agricultural areas particularly concerned by marginalisation processes. The abandonment of farm management under environmentally sensitive conditions of these areas does not only affect the agricultural sector, but might also imply negative effects for the whole regional economy and the viability of the socio-economic structure. The different mountain ranges have experienced quite contrasting development patterns over the past in this respect. Whereas for example over more than a century a number of valleys in the Western Alps have experienced a continuous abandonment of agricultural land use and emigration, the regions most threatened by abandonment have shifted nowadays towards the Eastern Austrian and Italian Alps. This trend is underpinned by the predomination of tourism centres in the more prosperous western parts of the Alps. It is the objective of this chapter to assess the role of rural amenities in mountain areas to nurture local development potential in order to cope with marginalisation threats.

With the shifts in Common Agricultural Policy (CAP) and the rising relevance of rural development issues, increasingly CAP's pillar 2 policy measures and the recent regional development programmes have addressed this potential. They reveal thus an example of policies addressing activities aiming at the provision of public goods and oriented towards overcoming marginalisation.

CHANGES IN REGIONAL DEVELOPMENT

Sustainable development in mountain areas has been a key concern worldwide since the early 1990s. The attention attached to mountain issues is increasingly related to the high ecological sensitivity of mountain areas and its impact on global change (Price, 1999). The inclusion of Chapter 13 'Managing Fragile Ecosystems: Sustainable Mountain Development' in the Agenda 21 document endorsed by the UN Conference on Environment and Development (UNCED, 1992) in Rio de Janeiro, is an indication of the priority of this issue. With the series of activities during the United Nations' International Year of Mountains 2002, awareness of these questions has been raised worldwide and the discussion has focused on the need for integrated policies to cope with the challenge of marginalisation in these areas.

The World Summit for Sustainable Development in Johannesburg (United Nations, 2002) reaffirmed the need to advance action on the issue and launched a cooperation framework, the Mountain Partnership, which is dedicated to deepen comparable analysis of mountain problems and prepare policy recommendations. One of the specific ongoing activities resulting from those discussions is the Sustainable Agriculture and Rural Development in Mountains (SARD-M) initiative, supported by the Food and Agriculture Organisation (FAO) and a great number of actors in mountain regions (FAO, 2004). It was acknowledged that improved policies and actions for the SARD-M initiative are urgently needed to meet the challenges of agriculture and rural development in mountain regions, where high levels of malnutrition and hunger persist, and to protect mountain environments for present and future generations, taking into account all the relevant factors. Although problems in mountain areas are not as serious in the European context as worldwide (Panos Institute, 2002), there are significant patches of mountain regions affected by marginalisation trends.

While there are numerous reports and analyses available for the development in the Alps (for example Tappeiner et al., 2003; Bätzing, 2002; Pfefferkorn and Musović, 2003), a comprehensive study on all the European mountain ranges has only been carried out just recently (Nordregio, 2004). It presents information on a wide set of topographical and socio-economic indicators and allows the tracing of changes at a low geographical level, that is, in general, the municipal level. The analyses include detailed information on demarcation options for mountain areas, and issues like land uses, demographic patterns and trends, on economic characterisation and the particular relevance of access, infrastructure and services in mountain areas. Figure 7.1 gives an overview of one of the core indicators for marginalisation, population density, and it reveals the low number of people present in many mountain areas of Europe, including the study area in Eastern Austria.

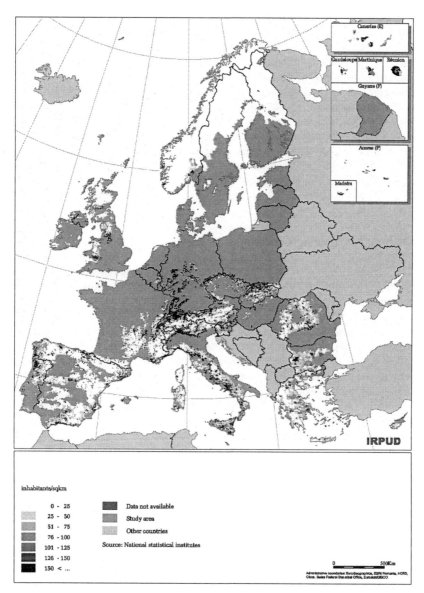

Source: Nordregio (2004: 77).

Figure 7.1   Population density in mountain municipalities

Analysis of population change over 1991–2001 by mountain municipalities moreover highlights the different processes of demographic change taking place in different parts of the European mountains. Whereas the general trend is stable or positive in Northern and Central Europe, in Eastern Europe and great parts of Southern Europe depopulation is the norm. Focusing on the Alpine mountain region one can realise that long-term population analysis is characterised by significant changes. Areas in the Western Alps where population decline took place for more than a century have now recovered and have been experiencing population increases over recent decades. On the other hand, great parts of the Eastern Alps have turned into regions with significant and persisting population losses. The study area of Neunkirchen is situated in this latter part of the Alps and thus can be taken as an example of recently increasing development problems (Wiesinger and Dax, 2005).

Study Area: Neunkirchen

The Neunkirchen district is situated on the eastern outskirts of the Alps, covering an area of about 1150 km² of which only a third is permanent settlement. The population amounts to 86 000 which results in a population density of 75 inhabitants per square kilometre, a level significantly below the national average of Austria. However, population distribution varies greatly within the area, ranging from above-average densities in the lowland parts to rather low levels in the mountains, and very thinly populated areas in some remote parts (down to only four persons per square kilometre in the most peripheral municipality).

While part of the north-east of the district consists of lowlands, the south and south-west are dominated by mountains and upland areas. The highest summit of the region can be found near the border to Styria (Schneeberg, 2076 metres). The proportion of forests and woodlands (mainly coniferous forests with pine, spruce and fir) in the land cover is extraordinarily high, exceeding the national average by far, particularly in the mountain parts. The lowlands of the north-east at an altitude of about 350 metres form the principal economic and settlement area, with the biggest towns and villages of the district. These are however small towns in the range of 6000 to 15 000 inhabitants which are heavily influenced by the nearby town of Wiener Neustadt (about 50 000 inhabitants) and the agglomeration of Vienna (at a distance of less than 100 km). The mountain areas are the locations for important summer and winter resorts which recall the first rise of popular tourism a century ago. They provide particularly interesting tourist facilities for day-trip visits from the agglomeration area of Vienna. As regards agricultural production, animal husbandry and dairy farming have traditionally been the most important sectors in the southern parts while arable land is more frequent in the north.

With regard to the topographical differences one can find quite distinct differences between the mountain areas (of the Semmering–Rax–Schneeberg Region in the west and south-west, and the Wechsel–Bucklige Welt region in the south-east) and the lowland part which comprises the main settlement and industrial area. To enhance the potential in the mountains a LEADER+ group (Niederösterreich Süd Alpin) has been set up for the Rural Development Programmes that are part of the second pillar of the CAP (period 2000–06). The analysis splits information for the two different parts of the region, thereby underpinning the relevance of detailed geographical analysis with regard to marginalisation issues in mountain areas.

In general this area enjoys high environmental quality in terms of pristine nature, cultural landscapes and biological diversity. Agricultural production is characterised by small and medium-sized structures and management systems are quite extensive. Thus environmental degradation as a matter of agricultural overexploitation is not of relevance in the region. On the other hand the consequences of extensification and abandonment of agricultural land use on landscape structure, and biological and agricultural diversity, might become more relevant. Since forestry is the overwhelmingly predominant land use, efforts towards more sustainable forest management are particularly important. Environmentally sound forestry proves an appropriate tool to prevent floods and other natural hazards in mountain areas. The area has also been the main water source for Vienna, supplying fresh spring water by pipeline for more than 100 years. Tourism is concentrated in some places (for example the Schneeberg–Rax region, winter tourism in Semmering and the Wechsel region), with moderate tourism intensity compared with the famous resorts in the high mountains of Tyrol or Salzburg.

The lowland part of the study area has traditionally been a core industrial area of Austria dominated by huge manufacturing plants and heavy industry. During the last 20–30 years many traditional industries have had to close down in the process of restructuring and privatisation. This has caused severe problems for the region in terms of rising unemployment, and more and more people were forced to commute (mostly to Vienna or Wiener Neustadt). In the last few years the economic decline has stopped and the industry of the region has recovered successfully through restructuring activities and the founding of a number of innovative and competitive small and medium-sized enterprises. Nevertheless the orientation towards centres outside the region has increased even over this period, stressing the fragility of the local economy. Although population development in the region was almost stable, there are significant local differences: population has declined by 15 per cent in the mountain region and by 40 per cent in the municipality of Schwarzau im Gebirge, the most remote part of the study area, over the last 30 years. As in many other Austrian regions overageing is starting to become an important issue. Already 26.3 per cent of the population is over 60 in the mountain region, while this reaches 22.7 per cent in the more favourable parts of the study area.

The outside orientation of the region had its impact on increased commuting and migration levels. For most municipalities, out-commuting is by far greater than in-commuting. Migration is an important issue particularly in the mountain part. For example, Schwarzau im Gebirge lost 11.3 per cent of its population within only three years (1999–2001) due to out-migration.

Table 7.1   Indicators for marginalisation

Indicator	Austria		Study area: Neunkirchen	
	Level/ change (%)	Marginal- isation relevance (%)	Level/ Change (%)	Marginal- isation relevance (%)
Population change, 1991–2001	+3.0	7.0	+0.3	16.5
Employment change, 1991–2001	+4.0	7.3	+6.1	5.5
Unemployment rate, 2001	5.1	0.9	5.5	0.9
Mountain difficulty, 2004 (average classification points)	68	10.2	103	15.0
Change of forest area, 1994–2004	+4.2	32.6	+1.9	7.9
Change of agriculture land use, 1994–2004	-7.1	61.6	-8.3	86.2

Note: Marginalisation relevance has been defined for municipalities fulfilling the following thresholds for the indicators:
1. population change in municipality (1991–2001) is lower than -5.0%,
2. employment change in municipalitiy (1991–2001) is lower than -5.0%,
3. unemployment rate in municipality (2001) is higher than 175% of Austrian average,
4. portion of mountain farms in 2004, as expressed through share in groups 3 and 4 (the two groups with highest production difficulties) – the system calculates the production difficulty for every mountain farm and can achieve a theoretical maximum of 570 points; groups 3 and 4 with most difficult production conditions have more than 180 and 270 points, respectively,
5. change of forest area in municipality (1994–2004) is higher than +5.0%,
6. reduction of utilised agriculture area in municipality (1994–2004) is stronger than -5.0%.

Source: Statistics Austria (2005), Bundesanstalt für Bergbauernfragen (BABF): own calculations.

This snapshot indicates that the negative trends of population losses in the mountain parts are still in place and continue to weaken the economic base of the local communities.

The following indicators of marginalisation processes compare the regional situation and trends with all municipalities of Austria (Dax and Wiesinger, 2005). The indicators are summarised in Table 7.1 with information both for the study area and the national situation, which is largely dominated by the situation in the Alps. The first column of Table 7.1 indicates the level of change, the second column the marginalisation relevance of each indicator. The interpretation is explained in the note below the table.

It is apparent that the share of municipalities affected by marginalisation trends is rather low when analysed by socio-economic indicators, but high for land use changes. This might reflect the different nature of the indicators and different time periods. Employment changes and migration patterns require a more general change in life prospects and thus refer to a rather long-term cause–effect relationship. On the other hand, land use changes are typically a direct result of the production potential in the sector, and largely influence the relative position of production with regard to other regions. However there might be a positive correlation between general economic performance and land use tendencies which could aggravate or mitigate the marginalisation threat in mountain areas. The differences between national and study area indicators call for studying the entire regional system in order to explain marginalisation processes. More detailed results include:

- Employment trends were rather positive over the analysed period and show minor differences between the two levels. One has to know, however, that the components of employment have changed considerably, leading to much higher shares of part-time employment, an increased employment rate for woman (which is still far below the men's employment rate) and the further separation between the place of residence and place of work. This adds to increased commuting areas, weakening the economic position of areas with predominant out-commuting.
- A highly negative population change is only experienced in some municipalities. Population loss is significantly higher in the study area, which underpins the perception of the local population of a region heavily threatened by economic problems.
- Indicators on agricultural production and mountain farming show that a large part of Austria, including the mountain part of the study area, is affected by severe production difficulties (handicaps). A detailed regional analysis on the situation all over Austria, of course, points to areas in Western Austria where this is even more clearly expressed.

- The indicators on forest area and agricultural land changes are related to one another. In almost all municipalities decrease in agricultural land is compensated for by an increase in forest area. For the study area itself the summary figures do not reveal the process sufficiently as the share of forest area is already so high that additional afforestation processes almost take up the last remaining areas not yet covered by forest. Under these circumstances, the figure on the decrease of agricultural land points more clearly to the marginalisation processes.

Overall, the analysis suggests that marginalisation in Austria is particularly affected by the country's difficult agricultural production situation in mountain areas. On the other hand, the socio-economic situation and positive economic performance holds true also when analysed at the regional level. Thus the areas affected by marginalisation processes are limited to rather small patches of the mountain areas in the country, a situation which is valid also for other parts of the Alps. Table 7.2 provides a calculation on land use changes and farm production difficulties. It underpins the findings that these tendencies are relevant in the mountain areas, and have to be taken into account when analysing marginalisation processes.

Table 7.2   Municipalities affected most strongly by land use changes, change of land use categories (1994–2004)

Indicator	Forest area increase, 1994–2004, more than +5.0% (1)		Agricultural land use decrease, 1994–2004, stronger than -5.0% (2)	
	Austria	Study areas	Austria	Study areas
Forest area change	+11.8	+11.1	+5.1	+2.2
Agricultural land use change	-9.5	-14.3	-12.6	-9.5
Mountain difficulty (average classifica-tion points)	80	188	98	130
Relevant area (%)	32.6	7.9	61.6	86.2
Overlays (1) and (2)	Austria: 24.6		Study area: 7.9	

Source: Statistics Austria, BMLFUW, BABF.

AGRICULTURAL ABANDONMENT AND AFFORESTATION

There is little evidence that agricultural abandonment is becoming more widespread in mountain areas than in other European regions. From literature on case studies (MacDonald et al., 2000; Baldock et al., 1996) and analyses of structural changes (Dax, 2004) it appears that tendencies of farm abandonment are particularly concentrated in the less fertile and drier zones in the Mediterranean, and in smaller areas at higher altitudes in various mountain chains of Europe, including medium-high mountains. Mountain areas represent the combination of socio-economic hardship and greatest physical constraints to agriculture, resulting in reduced economic viability of farming and in many cases regional development as well.

As the structural development suggests, there is not always a visible straightforward abandonment of agricultural land, but very often changes take place gradually and involve long-term adjustment strategies. Adjustment may be limited by traditional attitudes, inflexibility in production and fragmented structures, and if alternative uses are interesting a stepwise withdrawal might occur. These slow-changing patterns often conceal the long-term structural changes in land uses of mountain areas (MacDonald et al., 2000).

However, the opportunities for adjustment in farming are dependent on the competitive position of the rural economy and the comparative advantage for different types of economic activity. Baldock et al. (1996) have identified the particular vulnerability to marginalisation of small and extensive farming systems and the high relevance of such situations in mountain areas. Analysis of agricultural development in the Alps (Tappeiner et al., 2003) indicates that land abandonment has occurred particularly in the Western Alps, whereas in the Eastern Alps farm structures tend to be stable but farm management practices are shifting. There are many small regions affected by a continuous decrease in livestock numbers and a shift towards less-intensive schemes, with the study area being affected particularly by this trend. A continued minimum livestock density is essential in mountain areas to safeguard management of alpine pastures and to counter marginalisation trends. These include a large share of the UAA (utilised agricultural area) and are central to landscape development in high mountain areas. Alpine pastures also have a significant impact on biodiversity in Alpine areas. Livestock has gradually decreased during the last 20–30 years. This highlights the adjustment of farms to the changes in agricultural policy, particularly after Austria's European Union (EU) accession in 1995.

The other issue underpinning the vulnerability of farm management in the study area is the high proportion of forestry land use. With up to 90 per cent of forest area in some municipalities, agricultural land use become more or less secondary in these areas. From Figure 7.2 we can see that a similarly high portion of forest area is widespread in parts of the Eastern Alps in Austria, Italy and Slovenia, underpinning the tendencies towards marginalisation in these regions.

With regard to this situation, specific policy measures for mountain areas have been instituted, primarily to provide compensation for disadvantage. The CAP recognises the natural handicaps of the areas and their association with

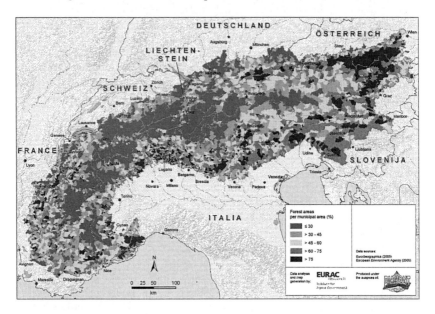

Source: Tappeiner et al. (2008).

Figure 7.2   Portion of forest land use in alpine area

depopulation and land abandonment through its structural support of the Less Favoured Areas (LFA) scheme since 1975 (Dax and Hellegers, 2000). Nowadays almost 56 per cent of the total UAA of the EU-25 is classified as LFAs, with the mountain areas attaining 17 per cent of the total UAA. Since the LFA support also includes 'other LFAs' outside the mountains, the widespread inclusion of new areas into the scheme and the scarce differentiation of support have been criticised by the Court of Auditors and led to a discussion on redefinition of LFA (Dax, 2005). In general only LFAs outside the mountains should be affected by the planned reform until 2010. However, it is important to underpin the permanent nature of the handicaps in mountain areas (and some other less-favoured areas) and the relevance of the social dimension for abandonment issues.

SHIFT IN TOURISM PATTERNS

Tourism is considered as one of the main alternative activities to farmers in mountain regions. Although it has been shown that in general the economic structure within mountain areas does not greatly differ from that in lowland rural areas, it remains true that tourism potential is particularly widespread and well known there. Yet it has to be acknowledged that there are considerable differences in tourism intensity between mountain ranges and within them. The example of the situation of the Alps shows that municipalities with the highest tourism intensity (more tourist beds available than inhabitants in the municipality) are concentrated in some valleys of the high mountains in the Western Alps (in France) and the Eastern Alps (in Austria). In contrast to the situation in France, tourism is of widespread relevance to the regions of large parts of Austrian mountains. As farm-based tourism is of particular relevance in Austria, the combination of income sources from agriculture, tourism and diversification activities is an important element against marginalisation trends. In this regard, although overall the intensity in the study area is limited (about nine beds per 100 inhabitants) it is an important sector within parts of the study area. The mountainous part is indeed famous for its summer resorts in the so-called 'belle époque' period at the turn of the nineteenth century. It was one of the first, well-renowned areas to spend holidays in the beautiful landscape at the outskirts of the Eastern Alps and in reach of the city of Vienna. After considerable structural changes it has regained importance for second homes and tourism patterns have particularly shifted towards short stays, in particular day-trip and weekend tourism. The number of overnight stays has declined considerably between 1994 and 2003 while the number of tourist arrivals has remained stable or even increased. This has led to a considerable reduction in the number of hotels, bed and breakfasts and overall beds available. The decrease of the local tourist industries has affected the financial situation of the municipalities and reduced the options of farm households for diversification linked to tourism.

A classification of intensive and less-intensive mountainous and non-mountainous municipalities for Austria shows the tremendous differences in intensity levels (Dax, 2004). Moreover, it reveals that in these mountain areas the regional processes in tourism development have tended for a long time towards further concentration (observation period 1975 to 1995): whereas the less-intensive municipalities (with an active population in tourism activities lower than 25 per cent) are characterised by a constant tourism intensity of about 20 overnight stays per year in relation to every inhabitant, this indicator attains about 160 for the more-intensive municipalities. The latter are, moreover, due to some further analyses on the recent development, very heterogeneous: a small sub-group of very intensive and expanding municipalities increased tourism performance by 2 per cent per year (1995–

2002) while the other municipalities showed stable or even declining results. This underlines the strong reliance on specific local or small-regional strategies, as they represent significant divergence between different groups of municipalities of a similar intensity. As in many of these areas activity rates in tourism-related economic activities exceed 25 per cent of the workforce, this sector's development has a direct impact on the overall regional economy, including farm–tourism cooperation potential. Moreover, given the lack of other alternatives for many other less-developed regions, the expansion and refocusing of tourism strategies was an attractive priority in regional development programmes.

It is important to notice also that tourism activities affects and transforms the natural environment of mountain areas. The calculation of the maximum persons present in the settlement area of mountain regions both underscores the ecological relevance of the intensity issue and clarifies that tourism is most influential at higher altitudes of the mountains of Western Austria. With 450 persons present per square kilometre of settlement area at peak periods, the density reaches a level comparable to the densely populated centres in the main valleys.

VALUATION OF RURAL AMENITIES

Agriculture plays an important role in maintaining multifunctional landscapes in many mountainous areas of Europe (Dax and Hovorka, 2004). In the Alps animal husbandry and grassland management are of major significance in the land use and decisive for landscape structures. Areas with a particular high nature value are widespread, as with high pastures, steep mountain meadows, dry grassland biotopes and damp meadows in some valleys sustained through extensive management systems. Mountain farms are also of great importance for forest protection and the management of Alpine pasture areas, which are extremely sensitive ecosystems.

The unfavourable natural conditions for mountain farming enterprises are expressed above all in the steep gradients of the farmed areas and the shorter growing season, being exacerbated by extreme weather conditions and implying an absence of alternative production possibilities. Often, an inadequate and expensive infrastructure, including high transportation costs and weak accessibility, may also be added to this. In many mountain regions farm holdings are moreover characterised by a small-farming structure which is operated primarily by family labour input. The average size of mountain farms in EU-15 is as low as 12.3 hectares UAA (against an average of 18.7 hectares UAA for all farms in EU-15). In terms of standard gross margin (SGM) the difference is even bigger: whereas the average SGM per holding in mountain areas is 8.1 Economic Size Units (ESU) this figure is up to 18.7 ESU for all the EU-15 farms (Dax, 2005). All these indicators refer to particular production difficulties and region-specific problems which have to be addressed through efforts to strengthen viability of agriculture in mountain areas.

The fact that only for 36 per cent of mountain farms in the Alps is agriculture the main economic activity has driven farmers towards the recognition of a wide range of functions, going far beyond mere food provision. Some of these wider tasks are linked directly to farming, but multifunctional mountain farming includes also objectives to sustain the management of externalities supplying services and values, reflecting a rising social demand (Crabtree et al., 2002). It is therefore important to take a comprehensive viewpoint of the roles of mountain agriculture in order to cope with development problems.

It seems important that under the difficult production situations of mountain areas the provision of these roles is linked to specific requirements of farm management with quite clear limits for intensification of production. Such production methods are particularly supported by the widely applied agri-environmental measures of the CAP. In this regard, the mountain farming strategies prioritising quality development and region-specific products represent a major asset and have a positive impact on farm household incomes. The activities deriving from such an approach reinforce the need for cooperation with other economic sectors and regional partners, and require observation of and orientation towards enlarged markets.

In the study area a particular valorisation of the resource base and an appreciation of the rural amenities has long been experienced. Starting from tourism development at the turn of the nineteenth century, the landscape of the region has been assessed as a specific target area for the Viennese population for short-term recreation within reach of the agglomeration. The images of 'pristine' nature have been underscored by the decision to provide the water supply for Vienna from the mountains of the region. This enforced the declaration of large areas as nature reserves and contributed to low intensity in land uses. These tight linkages to tourism and recreational uses have allowed farmers to manage agricultural land in this context despite production limitations for so long.

The underlying perspective only gained acceptance in the 1990s, when a shift from conservation to nurturing 'local assets' was realised for important parts of rural policy. The major aim of amenity policies is to exploit local resources and utilise local assets for rural development (OECD, 1999: 33), and to counteract thereby the vulnerability of these areas and the threat of marginalisation processes. It is made clear that the valorisation of amenities is the best incentive for preservation, but beyond this, the goal is to help rural territories harness the value of their amenities. Amenities have become a comparative advantage for some territories, like mountain areas, in part because amenities are highly specific to their location and cannot be transferred to other places like other assets. Some aspects of rural development policies in Europe have taken up the option to engage against abandonment of fragile rural areas, and have addressed mountain areas with specific instruments. The inclusion of mountain-specific objectives is most advanced in the policy sectors where a close link to the resource base and the amenity provision is visible (OECD, 1998). In particular, agricultural policy, rural development activities and regional policy include relevant measures and policy priorities (EC, 2003), all arguing to provide instruments against specific marginalisation processes. It is widely acknowledged that marginalisation of mountain areas cannot be tackled by measures of sector policies alone which refer to one problem dimension.

CONCLUSIONS

Many countries have realised the need to address the marginalisation threat of their mountain areas through elaborating, primarily sectoral, policies, laws and regulations for the use of resources. The underlying concepts are no longer so much based on considerations of preservation, but are increasingly inspired by higher valorisation and outside demand for the unique resources of mountain areas. Regional case studies and international work on rural amenity provision underpin the need to take account of the nature of rural amenities as public goods. In principle, they share common characteristics of uniqueness, irreversibility and uncertainty (OECD, 1999). As rural amenities are linked to the particular area in which they exist, mountain areas are characterised by specific sets of amenities which reveal a great local variety and are particularly valued because of the overwhelming diversity of their natural and cultural systems. However, market mechanisms tend to be unfavourable for these largely remote areas. In order to cope with marginalisation tendencies it is envisaged that combinations of market mechanisms and non-market approaches are required, particularly in remote areas. Regional development practices in mountain areas suggest that both an active core of local actors addressing the local market problems and harnessing the full development potential of the region, as well as the appropriate policy instruments, are essential to set up a significant development dynamic and withstand marginalisation.

An integrated approach seems necessary to take account of the multisectoral aspects of resource use and socio-economic development in mountain areas. Marginalisation can be understood as a multidimensional process encompassing not only land abandonment, environmental degradation and economic decay, but also social and cultural patterns. Agricultural marginalisation can hardly be unravelled from overall marginalisation in the regions, but it implies economic, environmental and social marginalisation. Land use change may be less apparent in the short term, but the slow trend of land abandonment and changes in use and intensity might have negative externalities in terms of biodiversity or other environmental issues in the mountains. There is a wide consensus reflected in CAP development that agricultural land use, particularly in mountain and other less-favoured areas, cannot be maintained by agricultural policies alone. Integration programmes combining regional, environmental, socio-cultural and economic development will also have to play a major role in combating marginalisation and land abandonment in these areas.

Moreover, as has been recognised, in order to provide a wide range of functions, going far beyond food production, a wide set of public goods is provided by agriculture which are particularly endangered in mountain areas. Particularly in this context, only the interrelation of these functions seems to provide a sound base to overcome the inherent economic problems and marginalisation threats. It therefore appears important to conceive policies and instruments which focus on the spatial aspects and limitations of the areas to establish viable farming structures. This view is strongly supported by the wider political context, including rising interests and demand from outside the mountain regions. Some of the instruments and mechanisms established in the CAP reforms (the LFA scheme, agri-environmental measures, cross-compliance, LEADER mainstreaming) and the structural funds implementation (mountain areas, recognition in spatial strategies, transnational cooperation) are examples of specifically addressing the mountain contexts. Key elements and principles for a policy approach to focus on sustainable development in mountain areas (Mountain Agenda, 2002) and to prevent marginalisation tendencies would be:

- recognition of mountain areas as specific development areas;
- remuneration for services rendered to surrounding lowland areas;
- diversification and exploitation of the local potential for innovation;
- cultural change without loss of identity;
- sustainable management of mountain ecosystems and biodiversity;
- taking account of spatial aspects to support cooperation and strategic approaches; and
- institutional development to focus on sustainable resource use.

Prospects for marginalisation are difficult to forecast since development in mountain areas is characterised by a large diversity and a great influence of local actors (Copus, 2004). Overall the quite serious trend of depopulation in large parts of European mountains will probably continue. The same holds true for mountain farming which is affected by market pressures and aspects of competitiveness, leading to specialisation and concentration of production. Up to now the CAP and rural development policy have provided existing farm structures with limited perspectives, shifting support only gradually to less-favoured and remote areas, like mountain areas (Arkleton Centre for

Rural Development Research, 2004; Shucksmith et al., 2005). However, the rural development approach applied in recent CAP reforms implies considerable potential to apply strategies favouring mountain areas. A more explicit differentiation of support according to production difficulties might better reflect the multifunctional nature and abandonment threat of mountain farming. Since low-intensity farming systems of mountain areas reveal characteristics which are to a high extent benign to the environment, but are endangered by both abandonment and intensification, there is an urgent need to highlight the importance of appropriate land management of these areas for landscape development, and support structures through appropriate policy programmes.

The high level of integration of the farming population in off-farm labour markets, pluriactivity and the role of regional policy for employment development underscores the second prerequisite for achieving regional objectives of sustainability and long-term provision of social demands. Through the provision of positive externalities, mountain farming contributes to maintaining settlement structure and shaping the cultural landscapes in areas which otherwise would lose significant parts of their development potential. Since by definition public goods are not rewarded in the market, there is an obvious case for transfers from society at large to reward those who maintain such public goods (Bryden et al., 2005). Thus the support for mountain farms is core for the positive direct and indirect effects in safeguarding the sensitive ecosystems and maintaining multifunctional landscapes in mountain regions. The debate on the socio-economic processes increasingly has to focus on the long-term provision of public environmental amenities to facilitate sustainable regional development and address the threats of land abandonment and marginalisation processes in mountain areas.

REFERENCES

Arkleton Centre for Rural Development Research (2004), 'The territorial impact of CAP and rural development policy', Final Report, ESPON Project 2.1.3, European Spatial Planning Observatory Network, Aberdeen. www.espon.lu and case study reports http://www.abdn.ac.uk/arkleton/publications/ESPONCaseReportsIndex.sh tml.

Baldock, D., G. Beaufoy, F. Brouwer and F. Godeschalk (1996), *Farming at the Margins, Abandonment or Redeployment of Agricultural Land in Europe*, London and The Hague: IEEP and LEI-DLO.

Bätzing, W. (2002), *Die aktuellen Veränderungen von Umwelt, Wirtschaft, Gesellschaft und Bevölkerung in den Alpen*, Berlin: Bundesministerium für Umwelt, Naturschutz, und Reaktorsicherheit.

Bryden, J., L. van Depoele and S. Espinosa (2005), 'Policies releasing the potential of mountain and remoter areas of Europe', Background Paper, Euromontana conference Reaping the Benefits of Europe's Precious Places, 9–11 November, Aviemore, Scotland.

Copus, A. (2004), 'Aspatial peripherality, innovation and the rural economy, Final Report', Aberdeen: Scottish Agricultural College.

Crabtree, R., N. Hanley and D. Macdonald (2002), 'Non-market benefits associated with mountain regions', Aberdeen: CJC Consulting.

Dax, T. (2004), 'The impact of EU policies on mountain development in Austria', paper at the Regional Studies Association International Conference, Europe at the Margins: EU Regional Policy, Peripherality and Rurality, 15–16 April, Angers, France, http://www.regional-studies-assoc.ac.uk/events/presentations04/dax.pdf.

Dax, T. (2005), 'The redefinition of Europe's Less Favoured Areas', paper at the 3rd Annual Conference, Rural Development in Europe: Funding European Rural Development in 2007–2013, 15–16 November, London.

Dax, T. and P. Hellegers (2000), 'Policies for Less-Favoured Areas', in F. Brouwer and P. Lowe (eds), *CAP Regimes and the European Countryside, Prospects for Integration between Agricultural, Regional and Environmental Policies*, Wallingford: CAB International, pp. 179–97.

Dax, T. and G. Hovorka (2004), 'Integrated rural development in mountain areas', in F. Brouwer (ed.), *Sustaining Agriculture and the Rural Environment: Governance, Policy and Multifunctionality*, Advances in Ecological Economics, Cheltenham, UK and Northampton, MA, USA: Edward Elgar, pp. 124–43.

Dax, T. and G. Wiesinger (2005), 'Analysis of marginalisation indices in Austria', EUROLAN, Wien: Bundesanstalt für Bergbauernfragen.

European Commission (EC) (2003), *Proceedings of the Conference 'Community Policies and Mountain Areas'*, 17–18 October 2002, Brussels: Commission of the European Communities, http://europa.eu.int/comm/regional_policy/sources/docconf/library/mountain_proceedings_en.pdf.

Food and Agriculture Organization of the United Nations (FAO) (2004), 'The SARD-M Project, strengthening mountain populations' livelihoods with improved policies for sustainable agriculture and rural development', Roma: Food and Agriculture Organisation, www.fao.org/sard/en/sardm/home/index.html.

MacDonald, D., J.R. Crabtree, G. Wiesinger, T. Dax, N. Stamou, P. Fleury, J. Gutierrez Lazpita and A. Gibon (2000), 'Agricultural abandonment in mountain areas of Europe: environmental consequences and policy response', *Journal of Environmental Management*, 59, 47–69.

Mountain Agenda (2002), *Mountains of the World, Sustainable Development in Mountain Areas, the Need for Adequate Policies and Instruments*, Bern: University of Bern, Centre for Development and Environment.

Nordregio (2004), *Mountain areas in Europe: Analysis of mountain areas in EU member states, acceding and other European countries*, Nordregio Report 2004:1 Stockholm, http://europe.eu.int/comm/regional_policy/sources/docgener/studies/study_en.htm

OECD (1998), *Rural Amenity in Austria, A Case Study of Cultural Landscape*, Paris: Organisation for Economic Co-operation and Development.

OECD (1999), *Cultivating Rural Amenities, An Economic Development Perspective*, Paris: Organisation for Economic Co-operation and Development.

Panos Institute (2002), *High Stakes: the Future of Mountain Societies*, Panos report no. 44, London.

Pfefferkorn, W. and Z. Musović (2003), 'Analysing the interrelationship between regional development and cultural landscape change in the Alps', Work Package 2 Report, REGALP, Regional Consulting, Wien.

Price, M. (1999), *Global Change in the Mountains*, New York and London: Parthenon Publishing Group.

Shucksmith, M., K.J. Thomson and D. Roberts (2005), *CAP and the Regions: The Territorial Impact of Common Agricultural Policy*, Wallingford: CABI Publishing.

Statistics Austria (2005), *Statistik Austria: Jahrbuch 2005* (Yearbook 2005), Bundesministerium für Land- und Forstwirtschaft, Umwelt und Wasserwirtschaft: Grüner Bericht 2005.

Tappeiner, U., A. Borsdorf and E. Tasser (2008), *Mapping the Alps*, Heidelberg: ©Spektrum Akademischer Verlag.

Tappeiner, U., G. Tappeiner, A. Hilbert and E. Mattanovich (eds) (2003), *The EU Agricultural Policy and the Environment, Evaluation of the Alpine Region*, Europäische Akademie Bozen, Berlin and Vienna: Blackwell Verlag.

United Nations (2002), 'Report of the World Summit on Sustainable Development', Johannesburg, South Africa, 26 August–4 September 2002, A/CONF.199/20*, New York.

United Nations Conference on Environment and Development (UNCED) (1992), Agenda 21, Chapter 13, 'Managing Fragile Ecosystems: Sustainable Mountain Development', Rio de Janeiro.

Wiesinger, G. and T. Dax (2005), 'Coping with marginalisation and multifunctional land use, Austrian case study Neunkirchen', EUROLAN report 2005/1, Wien: Bundesanstalt für Bergbauernfragen.

8. Marginalisation in Spanish dry areas: the case of Vilafáfila Lagoons Reserve

Jordi Rosell, Lourdes Viladomiu and Anna Zamora

The countries in the south of Europe present vast differences in several physical, climatological and productive conditions. There are areas with a very productive agriculture and capacity to export, but there are also large zones with very poor soil quality and considerable orographic difficulties, as well as low and irregular rainfall. The agrarian production in these zones has low rates of return. Agriculture in the south of Europe also includes very different farming systems. The intensive systems are in place on irrigated land dedicated to horticulture (vegetable and fruit crops), as well as other production such as sugar beet, alfalfa (lucerne), cereals, tobacco and cotton. Intensive husbandry systems of pigs, poultry, rabbits and calves complete the picture. Apart from modern and intensive systems there are traditional systems devoted to cereals, sunflowers, fallow and Mediterranean crops (vineyards, olive groves and almond trees) with sheep, goats and cattle present. The ecosystems found in the majority of Southern European regions are particularly fragile with high risks of erosion, fires and desertification. The population is very unevenly distributed, concentrated along the coast and in the big cities; the rest of the country has very low densities. In sum, a large part of the south of Europe can be classified as an agrarian area that is vulnerable to marginalisation due to topography, soil productivity, climatic conditions and socio-economic patterns.

In the current context of agrarian policy changes – the CAP reform of 2003 – and economic globalisation, the risk of marginalisation of some regions is returning as one of the issues on the European agenda. Spain was one of the countries that defended partial decoupling as a way to avoid land abandonment (MAPA, 2004).

The process of agricultural marginalisation may end in land abandonment. In Spain it is possible to differentiate between two types of areas that are more vulnerable to agrarian and land marginalisation: mountain areas and dry flatland areas.

The Spanish National Irrigation Plan estimates that around 7.8 million hectares of dry flatland, which consists of 30 per cent of the utilised agricultural area (UAA) in Spain, has no possibilities for irrigation. The Villafáfila Lagoons Reserve is an example of dry flatland oriented to cereal production. This agrarian system, very typical of the centre of Spain, is in ecological terms classified as a pseudo-steppe or a cultivated steppe (De Juana, 2005). This ecosystem plays a crucial role in the existence of steppe birds (Rosell and Viladomiu, 2005), especially the great bustard (*Otis tarda*), its most representative species. The area presents an additional interest due to a group of stationary saline lagoons that provides an attractive landscape in an arid surrounding.

Several studies have identified the risks and threats involved in the process of agrarian and land marginalisation of the steppe habitat (Alonso and Alonso, 1990; Díaz et al., 1993). Land abandonment, abandonment of traditional crops, land consolidation, cereal monoculture and increased use of chemical inputs have been pointed out as the main causes of the decline in the quality of the habitat. Agricultural production is positively related to the provision of habitat and food for bird species.

This chapter aims:

- To characterise a typical Spanish marginal agrarian area, considering both agrarian and rural parameters, and to define appropriate marginalisation indicators.
- To study the policies implemented in the area and their impact on the marginalisation process and on multifunctionality.
- To illustrate the problems of an area at risk of marginalisation in the context of decoupled payments, and to discuss the role of multifunctionality as a way to cope with marginalisation.

The chapter includes three sections. First, we describe the Villafáfila Lagoons Reserve in the context of key physical, social and economic indicators. Both quantitative and qualitative approaches are used. Second, we discuss the agrarian, development and conservation policies implemented in the region and their impact on marginalisation. And finally, in the conclusion, we include the most appropriate measures to limit the risks of agricultural marginalisation and land abandonment considering the multifunctional capacity of agriculture to cope with marginalisation.

FEATURES OF THE CASE STUDY AREA

General Characteristics

The Villafáfila Lagoons Reserve case study includes 389 km^2 in 11 municipalities, located in the Zamora province of the Castilla y León region. Considering the European structural funds status, Castilla y León is an Objective One region. The 11 municipalities are included in the agrarian county of Campos-Pan (Figure 8.1).

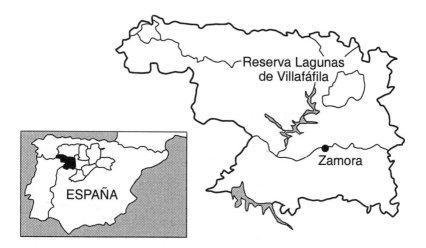

Figure 8.1   Villafáfila Lagoons Reserve

The list of the marginalisation indicators for the Villafáfila case study is presented in Table 8.1 with their interpretation in the Spanish dry flat region context.

The area was declared a National Hunting Reserve in 1986 and one year later it was put on the list of Special Protection Areas for Birds due to the exceptional wealth and diversity of the reserve's bird life (Directive 79/409/EEC). Several of the bird species that can be found in this zone are identified as being under varying degrees of threat of extinction (Birdlife International, 2000). The series of seasonal saline lagoons in the area have been included in the Ramsar Convention's List of Wetlands of International Importance since 1989. It is expected that the regional administration will declare it as a Nature Reserve by 2010. The Villafáfila Lagoons Reserve is part of the Natura 2000 network.

Table 8.1   Indicators and their interpretation

Topic	Indicator	Study area level	Interpretation and thresholds
Climate	Precipitation	Semi-arid and no irrigation	Limits yields and alternatives. The study area is part of the 30% of the Spanish UAA with no possibilities to be irrigated
Topography	Altitude	600 m	Contrasts in temperature do not allow for the most valuable Mediterranean crops
Soil quality	Rotational model – traditional fallow	40% of fallow	Performances of agriculture is reduced by fallow
Yields	• CAP reference yields • Minimum yield	2500 kg/ha Estimated at 1800 kg/ha	Reference level is below the national average and close to the estimated minimum
Farm size	AWU per farm	0.74	Farm size not enough to assume work for a person per farm
Pluriactivity	Full-time farmers	88%	Pluriactivity is an exception and far below the national average
Population	• Density • Population aged over 64 years • Structure of population agglomeration	9.76 inhabitants per km^2 37% of the population over 65 years old All villages have less than 1000 inhabitants	Minimum density at county level to maintain social services network (10 inhabitants/km^2) 16% of the Spanish population over 64 Lack of urban hierarchy
Gender structure	Men per woman in the age group 20–45	0.74 women per man	Gender imbalance indicates demographic difficulties
Labour market	• Activity rate • Agrarian employment	38% (women 20%) 43% agrarian employment	Very low level of women's activity rate High agrarian specialisation
Income	Agrarian and non-agrarian public transfers	Around 70%	High dependency on CAP payments and retirement pensions
Accessibility	Distance to the main economic axis of Spain (km)	250 km away from Madrid	Peripherical situation at Spanish and European level
Social capital	Main associations	No important non-farmers associations	Control of the institutional networks by farmers

The climate of the area is continental Mediterranean with sharp temperature changes, very cold winters and dry hot summers (from an annual average temperature for the coldest month of minus 1.3°C to an average maximum for the hottest month of 33.8°C). There are frequent frosts in winter that sometimes extend into spring. Rainfall (annual average rainfall is 431.9 mm) is irregular and scarce. Strong winds are a fairly common feature. The precipitation regime and the frequent frosts limit crop alternatives to a small group of products and imply low and variable yields. Although there are very large underground water resources, the water is extremely brackish and not fit for irrigation or human consumption.

The area is flat, it rises up nearly 700 metres, and is characterised by a gentle undulating relief. Altitude operates as a handicap for agriculture development. Land consolidation efforts in the 1970s entailed the disappearance of virtually all the tree cover in the area and facilitated agrarian mechanisation. Natural vegetation in the form of trees and bushes is currently very scarce and limited to the banks of the watercourses and the outskirts of the towns.

Agrarian crops are grown on a rotation system due to low soil quality and lack of irrigation. The traditional rotational model of production in the area implied a significantly high proportion under fallow. Up to 50 per cent of the land is left fallow, meaning that during a normal year only half of the land is productive. During the 1960s and 1970s, use of fertilisers allowed the reduction of the amount of land under fallow. These changes have generated the so-called 'intensified alternative' followed by most farmers, as opposed to the 'traditional alternative' (Díaz et al., 1993). The CAP reform of 1992 (Viladomiu, 1994) reinforced the cereal monoculture tendency, increased agrarian land, introduced a very complex fallow and set-aside system, and motivated temporary crop changes. The agri-environmental schemes and, to a lesser extent the LIFE programme, were very important for maintaining alfalfa and other pulse crops.

Agricultural land in the study area accounts for 90 per cent of total land. Pastures and woods represent 5 per cent and 0.4 per cent, respectively. Cereals (32 per cent of the cultivated area in 2004) and fallow land (39 per cent) account for around three-quarters of the UAA. This level of fallow is highly influenced by the CAP and the set-aside regime. Other crops are alfalfa, sunflowers and other pulses. The comparison of the Agrarian Censuses of 1989 and 1999 shows that arable land increased by 1.3 per cent in Villafáfila Lagoons Reserve.

If we compare the reference yields (2500 kg per hectare) for cereals, the area has lower levels than the Spanish average and is very far from those of the main production areas with high yields in Germany, the Netherlands and France, whose levels are nearly three times the yield levels of the selected area. The threshold is nowadays around 1800 kg per hectare for winter cereals. If we consider that almost a quarter of the land is under fallow and exempted from direct aid, the production level of the area is around this yield minimum. For most of the farms, their positive standard gross margin (SGM) is due to the direct payments received.

The number of farms is decreasing. In Villafáfila Lagoons Reserve, about a quarter of the farms disappeared during the period between 1989 and 1999. This decline indicates a sharp adjustment process. Farm size is increasing. Data from the Agrarian Census of 1999 indicate an average size of 51 hectares per farm with 13 plots and 0.74 annual work units (AWU)[1] per farm. Most of the farms do not reach one AWU or the equivalent work performed by one full-time person per year. Farm pluriactivity is not very developed in Villafáfila Lagoons. The Agrarian Census of 1999 shows that 88 per cent of the farmers and their spouses are mainly full-time farmers[2] (compared to the Spanish average of 67 per cent). In addition, 79 per cent of the other family members spend more than 50 per cent of their working time on the farm. The low level of pluriactivity in Villafáfila Lagoons Reserve is not due to the structural characteristics of farms. It is more related to other economic factors, such as the lack of external jobs and opportunities.

Population and Employment

The area underwent severe depopulation during the second half of the twentieth century, and this process took place in most villages and small towns in the Zamora province (Rosell et al., 2000). The migration process was a consequence of the mechanisation of the countryside and the increase of urban labour demand. During the 1990s, out-migration has decreased but the population is still declining (Table 8.2) due to a higher mortality rate than birth rate.

During the period 1960–2005, the number of inhabitants decreased from 10 216 to 3784. Nowadays, the population density is less than ten inhabitants per square kilometre. This density is often considered by the public administrations as the minimum level required to maintain the adequate social services network.

The inhabitants live in villages; there is no dispersed population. Municipalities are demographically similar (none have more than 1000 inhabitants) and there is no town that plays the role of county centre, concentrating public and private services. Several public and private services have left and others are vulnerable to closing down due to the lack of critical mass in all the towns of Villafáfila Lagoons Reserve.

Lack of employment and opportunities for women has meant their exodus from the area. As a consequence, the structure of the population shows a higher proportion of men, except in the age group over 65 years. Particularly serious is the imbalance between genders in the age group between 20 and 45 years (where there are 0.74 women per man) (INE, 2004).

Table 8.2   *Evolution of the population in the 11 municipalities of the selected area (1845–2005)*

	1845–1850	1960	1991	2005
Cañizo	462	807	392	306
Cerecinos de Campos	818	1 205	460	396
Manganeses de la Lampreana	732	1 718	920	689
Revellinos	352	724	342	299
San Agustín del Pozo	116	413	195	211
San Martín de Valderaduey	386	324	101	87
Tapioles	480	447	216	215
Villafáfila	1 260	1 783	623	619
Villalba de la Lampreana	426	717	322	289
Villárdiga	255	365	123	99
Villarrín de Campos	854	1 713	551	574
Total population	6 141	10 216	4 245	3 784
Population density (inhabitants/km^2)	15.8	26.3	10.9	9.76

Source: INE (Instituto Nacional de Estadistica/National Statistics Institute), Population Censuses (several years).

In 2001, the activity rate[3] was only 38 per cent in the study area, 50 per cent in the Castilla y Léon region and 56 per cent in Spain. This low rate is mainly due to the low level of women's participation in the workforce and the large proportion of elderly people. The female activity rate is around 20 per cent in the study area, 37 per cent in Castilla y León and 44 per cent in Spain (INE, 2003). During the 1990s the female activity rate of the study area has increased by 5 per cent, whereas the male activity has decreased by the same percentage. The fall of the male activity rate is due to both a longer time spent in the education system and the ageing of the population.

Total employment in the area has been decreasing. During the 1990s, employment was reduced by 6 per cent in Villafáfila Lagoons Reserve, whilst in Spain employment increased by 31 per cent. The unemployment rate for the 11 municipalities of the Villafáfila Lagoons Reserve was 13 per cent in

2001. The female unemployment rate (18 per cent) was twice that of the male unemployment rate (9 per cent). A high unemployment rate and a low activity rate illustrate social difficulties in the area. The economy of Villafáfila Lagoons Reserve is mainly based on agriculture and livestock farming. In 2001 the agriculture sector employed 43 per cent of total labour force, compared to 9.2 per cent in Castilla y León and 5.9 per cent in Spain (Table 8.3). The agrarian sector declined during the 1990s, and the service sector has increased. However, agriculture remains the main employer.

Table 8.3   Employment in Villafáfila Lagoons Reserve by sector (total number and percentage)

	Agriculture (%)	Industry (%)	Services (%)	Total
1991	730 (56.7)	202 (15.7)	356 (27.6)	1288
2001	518 (42.8)	205 (17.0)	487 (40.2)	1210

Source: INE (Instituto National de Estadistica/National Statistics Institute), Population Censuses (several years).

Over the period 1991–2001, employment in the agriculture sector decreased by 30 per cent, and the increase in service employment was not enough to compensate for the agrarian jobs lost. Taking these data into account, the Villafáfila Lagoons Reserve area has a very high level of dependence on agriculture and the agri-food sector. In addition to 43 per cent of direct agrarian employment, we estimate that half of the industry employment is in the food industry and 10 per cent of the services are directly related to agriculture. In total around 55 per cent of employment is due to agrarian activity.

Income and Accessibility

We use provincial data as a proxy due to the lack of information at the municipal level. In 1999, net farm income per agrarian employed and taxable income per employed in Zamora province was around 90 per cent of the Spanish average. The agrarian income per agrarian employed is about 55 per cent of the taxable income per employed in both Zamora and Spain (Table 8.4).

Agriculture in the study area has a high dependency on farm support payments. Direct payments are around 44 per cent of the Zamora agrarian income (Junta de Castilla y León, 2002). There are no data for the study area, but due to its high level of specialisation in arable crops and the importance of the agri-environmental schemes, the result could be an even higher dependency level. Although statistics are not available, it is worth highlighting the importance of retirement pensions and direct payments in total income. It is likely that pensions account for more than a quarter of the income. The percentage of the population over 64 is more than twice the Spanish average (37 per cent versus 17 per cent, respectively). If agricultural payments are added to this, over 70 per cent of the income in the area is accounted for by public transfer payments. The area is very dependent on public payments and very vulnerable to changes in policies.

Table 8.4   Net farm income and taxable income (1999)

	Net farm income (thousand €)	Agrarian employment (thousand people)	Net farm income per agrarian employed
Zamora	257 001	15.7	16 369.49
Spain			17 991.50
	Taxable income (thousand €)	Total employment (thousand people)	Taxable income per employed
Zamora	1 940 332	64.9	29 897.71
Spain			33 107.07

Source: INE (2002).

Zamora province is one of the most peripheral provinces of Spain. Being on the border with Portugal and having poor links with other Spanish regions has traditionally given it the characteristic of being a cul-de-sac. The area (taking Villafáfila as the reference point) is 23 km from Benavente, a commercial centre, 40 km from Zamora city (the provincial capital of Zamora), 100 km from Valladolid (the regional capital of Castilla y León) and 250 km from Madrid.

Road transportation networks have been markedly improved over the past few years. In 1985 Zamora had a road network of only 2500 km. Ten years later there were 4000 km of roads. In 1980, the province of Zamora had neither dual carriageways nor highways. Twenty years later, there were about 100 km of dual carriageways. These dual carriageways connect Madrid with Galicia, Madrid with Asturias, and Valladolid with Zamora city. Main roads are in good condition. However, further improvements are needed, mainly in upgrading secondary roads. Improvements are especially needed with the connection between Villafáfila and the A6 dual carriageway (Madrid–Galicia). The railway line of the study area is out of use. Zamora does not have an airport.

The telephone network and Internet coverage via phone or satellite are complete. The time needed to access some basic services by private car has been measured (see Table 8.5). Special public transport exists for students. There are social security agreements in order to cover transport expenses for sick persons. The general public transport is limited to three services twice a day that connect some of the municipalities of the area with Zamora.

Table 8.5   Accessibility to centres and services

	Town / village	Time (minutes)
Schools	Almost all municipalities	< 15
High schools	Villalpando	30
University colleges for undergraduate students	Zamora	45
Universities	Salamanca and Valladolid	90
Primary sanitary services	Almost all municipalities	< 15
Hospitals and specialists	Zamora	45

In conclusion, the peripheral character of Zamora influenced the development of the Villafáfila Lagoons Reserve. Nevertheless accessibility to centres and services are considered to be acceptable.

Social Capital

Social capital refers to the institutions, relationships and norms that organise the social life of individuals, and the many and varied ways in which a given community's members interact (Putnam, 1993). In a given region, social capital is based on abstract elements such as mutual trust, network capability, social cohesion, entrepreneurship and the cooperation between public and private sectors. Increasing evidence shows that social capital is critical for societies to prosper economically (Stulhofer, 2000).

In Villafáfila Lagoons Reserve, town councils are the most important administrative and political bodies. However, they are not large enough and there is little cooperation amongst themselves. Despite this, they have promoted projects that have had an impact on the quality of life (parks, sport centres) and on tourism development (a golf course).

The main farmers' unions actively approach public authorities to lobby them. They do not have any territorial development project. In neighbouring districts there are a number of farmers' cooperatives marketing livestock products, such as sheep milk, and cereals. There are few voluntary organisations, although arts and/or leisure groups have been formed.

Low productive diversification of the area, highly protected agriculture and an elderly population are elements that negatively affect the development of the social capital of the area.

The self-confidence of the local agents is limited. They are mainly farmers and they expect the public administration to solve their problems by means of intervention prices, direct payments and/or with the provision of services and infrastructure. The tendency to innovate has been low and the entrepreneurial spirit reduced. In fact, the general trend among farmers has been, for decades, to send their children out of the area in order to find work outside of agriculture. This aspect reveals the strong pessimism of farmers about the future of the agrarian sector.

THE IMPACT OF POLICIES ON MARGINALISATION AND MULTIFUNCTIONALITY

Agricultural Policy

Due to its cereal and ovine specialisation, agrarian policy is basic in the Villafáfila Lagoons Reserve. Thus, the 1992 CAP reform had an important impact. Non-irrigated alfalfa and other pulses did not receive compensatory payments and could not compete with cereals. The tendency for cereal monoproduction was reinforced. This process was strengthened by the intensification of the ovine production system and the difficulties in the mechanisation of some pulses. During the early 1990s, alfalfa and pulse production decreased enormously.

The set-aside payment introduced in the CAP reform of 1992 forced the development of the 'white fallow' concept in Spain. This was done so as to differentiate between traditional fallow and the new set-aside. The land included under the white fallow does not receive compensatory payments. Due to the difficulties in establishing individual levels of white fallow, in the middle of the 1990s the national Ministry of Agriculture approved different levels of fallow in each county. For the study area the level is 23 per cent. In addition, as a result of the direct payments regime of the CAP, the farms of the area where production exceeds 90 tonnes (or 37 hectares) are obliged to put another 10 per cent aside in order to be eligible for the payments. Another proportion of the fallow is the result of the voluntary set-aside, which accounted for 10 per cent of the arable land most years and, as an exception, 20 per cent.

Almost every year, voluntary set-aside has been at its maximum level. The voluntary set-aside indicator informs us about the low level of soil quality and low profitability. Several surveys from the Ministry of Agriculture use this indicator to assess the Spanish regions with a high future risk of land abandonment under a decoupling payments regime. During the period

2000–04, with a payment of €157.50 per hectare for voluntary set-aside, farmers excluded from production a minimum of 40 per cent of the agrarian land. In other words the standard margins do not exceed set-aside payments for at least 40 per cent of the land.

The reduction of alfalfa and other pulse production endangered the great bustard (*Otis tarda*) and other steppe birds by reducing their food supply. It is estimated that alfalfa should represent at least 8 per cent of the total area in order to guarantee an ideal habitat for the great bustard (Alonso and Alonso, 1990).

From 1992 the economic results of the Villafáfila Lagoons Reserve farms have been positive, mostly owing to direct payments from the first pillar of the CAP and compensatory payments from the second pillar of the CAP. The area is eligible for compensatory allowances as an EU less-favoured area (Directive 75/268 EEC) under high risk of depopulation. As we pointed out in the previous section, we can estimate that direct payments represent around half of the farm income.

Development Policies

Villafáfila Lagoons Reserve is covered by a number of measures, programmes and policies that affect its economic and social development. The design, funding and implementation of the policies affecting the area involve many different (semi-)public institutions at various levels. The European Commission, national government, regional government of Castilla y León, the provincial administration of Zamora (Diputación) and town council are the main agencies involved in the development of the area through various measures, programmes and policies.

The Villafáfila Lagoons Reserve is part of an area with a low level of development. Various measures and programmes aim to accelerate development and close the gap with the more highly developed regions of Spain and Europe. Funding has been channelled through the Objective One programme for the period 2000–06.

As in all of Spain, there is a distinct bias in the Castilla y León Objective One programme towards funding transport and communication infrastructure. As we pointed out, a significant improvement of the road system has been carried out during the period 1990–2005.

Most of the agriculture and rural development budget of the operational programme is oriented towards irrigation and land consolidation projects. The Villafáfila Lagoons Reserve has not benefited from them. This is largely because the consolidation works were implemented a long time ago and the irrigation projects are incompatible with the conservation of the natural habitat.

The Villafáfila Lagoons Reserve is included in the area covered by one of the LEADER+ groups, Adri-Palomares. The local action group covers 1998 km^2 in a region with some 23 500 inhabitants. One of the most relevant projects is the Visitors Centre of Villafáfila Lagoons opened in September 1995, the cost of which was borne by the public sector. The opening of the centre was to help draw attention to the steppe's ornithological richness.

The development policies have had a significant impact only in the indicator of accessibility. Other indicators are affected by the new labour opportunities thank to the LEADER programme, social projects and the Visitors Centre of Villafáfila Lagoons.

Conservation Policies

Compensation for damage

The study area was designated as a National Hunting Reserve in 1986. By that time, the regional government had introduced a wildlife crop damage compensation scheme for those farms affected by the reserve measures. The scheme is based on technical assessment carried out by staff from the regional Ministry of Environment, and specialists appointed by the local governments (municipalities) at the request of affected farmers. The aim was to establish the amount to compensate farmers for the damage produced by fauna, mainly birds. The main criticism of this scheme is its bureaucratic complexity. In 2005, the municipalities bordering the Villafáfila Lagoons Reserve have requested payments from the same scheme. In their opinion, they are suffering similar damages due to the increased fauna.

Agri-environment policies

Agri-environmental measures have acquired a notable relevance in the zone. The first agri-environmental programme applied in the area was the Cereal Steppes Scheme (CSS). This programme was promoted by the Consejería de Medioambiente with the collaboration of a conservation association, the Sociedad Española de Ornitología (SEO) (a member of Birdlife International).

The CSS was developed during the 1990s under the agri-environmental programmes, to maintain the habitat quality of steppe birds and to protect them from the risks of land use changes or land abandonment. During the early stages of the CSS, the uptake level was very low (Petersen, 1998); in 1996, approximately 3.8 per cent of the dry land of the reserve participated in the scheme (Rosell and Viladomiu, 1996). However, thereafter the level of involvement grew substantially to the point of accounting for almost all farmers who were eligible. The programme proved useful in increasing and improving alfalfa cultivation within other environmental measures.

In 2001, the Protection Programme of Flora and Fauna in extensive agriculture areas (SEPFF) replaced the CSS. The programme remained largely unchanged, but the payments were reduced from €72 per hectare for the CSS to €56 per hectare on average in 2004. In addition, the annual budget of the programme is much lower; therefore, the number of hectares that can benefit from it is also lower. The programme does not include specific support for alfalfa-growing land.

In order to tackle the risk of a decrease in alfalfa-growing land, the regional Ministry of Environment approved a specific programme in 2002, the Non-irrigated Alfalfa Scheme. This programme awards a bonus of €427 per hectare for alfalfa-growing areas. From the point of view of farmers it is a good scheme, with extraordinarily high bonuses. But due to budgetary restrictions only a limited area can benefit from this programme. Farms located in the area of the Villafáfila Lagoons Reserve have priority. The Ministry manages the programme and selects the plots of land following the criterion of bird density. The number of requests was much higher than the number of beneficiaries. Only 1900 hectares were selected out of 8000 hectares. Farmers do not agree with the system used in determining the share-out. Finally, a programme called the Non-irrigated Sunflower Scheme was created to support this crop when the direct payments for herbaceous crops, such as sunflowers, decreased.

LIFE-Nature programmes

Three programmes were implemented, and two of them aimed to preserve the great bustard's habitat by means of the purchase of land by the regional government in order to grow alfalfa. A third programme aimed to fight against the disappearance of nesting sites for birds by the renewal of dovecotes.

Policy Impact on Marginalisation and Multifunctionality

Looking at the policies implemented in the Villafáfila Lagoons Reserve, the most outstanding features are the high percentage of land under agri-environmental schemes. Table 8.6 shows that 61.6 per cent of the cultivated area falls under such programmes.

The agri-environmental programmes are positively evaluated by farmers as well as by technicians of the administration and experts (Atance and Barreiro, 2004):

- First of all, agri-environmental programmes contribute to the maintenance of cultivated land as they increase direct payments. In this sense, agri-environmental schemes have avoided land abandonment.
- Secondly, agri-environmental schemes have contributed to the maintenance and improvement of the quality of steppe bird habitats and have contributed to the recovery of alfalfa growing. Although the environmental impact of the agri-environmental schemes has not yet been appraised, it stands out that these, together with the two LIFE-Nature programmes, have contributed to the increase of the alfalfa-growing area.

Table 8.6   Involvement in agri-environmental schemes in the Villafáfila Lagoons Reserve (ha, 2004) and share in total cultivated land (%)

Scheme	Involvement
Cereal Steppes Scheme (CSS)	4 196 (11.8)
Non-irrigated alfalfa	1 877 (5.3)
Non-irrigated sunflower	96 (0.3)
Protection programme of flora and fauna in extensive agriculture areas	15 761 (44.3)
Total agri-environmental schemes	21 931 (61.6)
Cultivated land	35 609

Source: Adapted by the authors from Junta de Castilla y León data.

As a result of all these measures, the area growing alfalfa increased from 1455 hectares in 1997 to 2861 hectares in 2001. In 2004 the area continued to increase and stood at 3575 hectares. The districts of the Villafáfila Lagoons Reserve that have the highest percentage of alfalfa growing land are those that have the highest population of livestock. However, the changes in the ovine food system towards semi-stabling negatively affect the areas growing forage and alfalfa. This limits the synergies between agriculture and livestock. Some of the interviewees pointed out that the purchase of alfalfa by the dehydrated feed industry has also contributed to the recovery of alfalfa over recent years.

• Thirdly, it is thought that agri-environmental schemes have contributed to a change in the farmers' opinion of the steppe's ornithological richness. Farmers no longer see this richness as a handicap for the development of their activity, but as an economic reward.

Another element to be considered is the importance of the Visitors Centre of Villafáfila Lagoons Reserve in attracting tourism and contributing to the popularity of the area. During the years 1997–98, the number of visitors was over 50 000 people. From 1998 until 2003, the number dropped due to the lack of promotion (Table 8.7).

The impact of such a number of visitors was considerable during the early years, especially in Villafáfila, where new restaurants, bars and accommodation were created with the support of the LEADER programme. In fact, the LEADER programme, through Leader II (1994–99) and LEADER+ (2000–06), has contributed to the creation of an incipient rural accommodation network in the area, some new cafes and restaurants, and the completion of the Villarrín golf course. The construction of a new visitor's centre devoted to dovecotes is the latest attempt to attract tourists.

*Table 8.7   Number of visitors to the Visitors Centre of Villafáfila Lagoons
              Reserve*

Year	Visitors
1995	11 017
1996	44 393
1997	57 808
1998	57 729
1999	46 005
2000	46 597
2001	44 407
2002	31 000
2003	35 182

Source: Adapted by the authors from Junta de Castilla y León data.

Beyond some rural and nature tourism, the productive diversification of
the area has remained limited. The LEADER programme has contributed to
the growth and modernisation of the local cheese industries in Tapioles and
Villarín, as well as the establishment of the same industry in Manganeses de
la Lampreana. The programme has also assisted in the construction of a
sausage factory in Villafáfila, and has assisted in refurnishing a textile
cooperative in the same village. Nevertheless, the development of the
manufacturing sector is very slow, having lost certain industries, such as the
Villarín bakery. Agrarian diversification in the area is limited to the creation
of a small amount of rural accommodation for tourism.

The quality of life in Villafáfila Lagoons has been improved by other
public projects such as the renewal of dovecotes with LEADER and LIFE
resources, the construction of social and sports facilities from the regional
government budget, the implementation of a waste management centre, and
public space improvements in most of the villages.

CONCLUSIONS

The selected indicators illustrate the vulnerability of the Villafáfila Lagoons
Reserve towards marginalisation and the risk of land abandonment.
Meteorological characteristics and lack of irrigation limit crop alternatives.
Poor soil quality forces the practice of a rotational productive system with
around half of the land under fallow. The yield of the dry land in the

Villafáfila Lagoons Reserve is very close to the minimum considered as profitable. For most of the farms, their current positive SGM is due to the direct payments received.

The density of population of Villafáfila is around ten inhabitants per square kilometre. Several programmes and projects implemented by the Ministry of Agriculture have established this level as the minimum threshold to sustain a social service network at county level. The loss of population correlates to the evolution of the labour market and agrarian specialisation. The lack of female jobs means a greater emigration of women and, consequently, a gender imbalance in favour of men within the active population group. In the Villafáfila Lagoons Reserve, the age bracket between 20 and 45 years old is characterised by one man for every 0.75 women. Moreover, the ageing of population is one of the main concerns in Villafáfila, with 37 per cent of its population being over 65 years old.

The Villafáfila Lagoons Reserve is located far from Spain's most dynamic economic regions. The associative sector is not very developed. Due to the importance of farm unions, the associative organisations are mainly oriented towards lobbying activities instead of local project development efforts. The region depends on agrarian and non-agrarian public transfers. Economic diversification is very limited and the number of farms is sharply decreasing. Pluriactivity is exceptional and the CAP plays a crucial role in farm strategies.

The Villafáfila Lagoons Reserve is an area which is vulnerable to marginalisation, taking into account agrarian and rural indicators. The agrarian fragility is well illustrated by the indicators of climate, lack of irrigation, soil quality, fallow land and yields. Farm size and pluriactivity indicate farm vulnerability. A low population density, gender structure, agrarian dependency and employment and unemployment rates show the rural decline of the area. Finally, income transfer dependency and a poor social capital indicate the risks of the study area.

Over the last few years some measures implemented in the Villafáfila Lagoons Reserve have contributed towards improving the situation. The degree of isolation is decreasing due to improvements in the road network. Health, education and population services are improving. The renewal and construction of houses are common in all municipalities. Due to the developments in tourism, motivated by ornithological interest, the Villafáfila Lagoons Reserve has new restaurants, cafes, accommodation services and craft shops. Town councils and the LEADER group have played an important role in supporting new economic activities. Some indicators, such as the number of second homes and the seasonal fluctuation of population, indicate a better social appraisal of the Villafáfila environment and an improvement in the quality of life.

Other policies are contributing towards increasing income levels, mainly due to the 1992 CAP reform and other social transfers (unemployment subsidies, retirement pensions, and so on). However, these policies have also triggered an increase in the economic dependency of the Villafáfila population on income transfers. The Villafáfila Lagoons Reserve is a good example of the problems in central dry areas of Spain; it shares the same characteristics with 30 per cent of the UAA in the country with no possibilities of being irrigated. The indicators used to illustrate marginalisation are common in most of the counties of the dry areas of the Spanish central plateau. Only the main villages and some irrigated areas show a different socio-economic trend. Some features of the case study are:

- Agriculture does not only produce food and fibre. The pseudo-steppe ecosystem is the basis for maintaining ornithological richness and the picturesque landscape of the lagoons.
- In comparison to other regions, the Villafáfila Lagoons Reserve has exceptional expertise in the application of programmes and schemes to maintain agrarian multifunctionality. The approval of a Hunting Reserve introduced the damage compensatory scheme. At the same time, around 62 per cent of the agrarian land received benefits from the agri-environmental measures and three LIFE programmes were implemented on the reserve.
- The CAP reform from 2003 remains a challenge for the region. The MacSharry reform from 1992 reinforced the tendency towards cereal monoproduction with a negative impact on the maintenance of ornithological richness, but avoided land abandonment.
- The reform from 2003 with decoupled payments – fully or partially – may be a system that stimulates land abandonment. In the opinion of agrarian technicians, the low standard gross margins indicate a high risk of land abandonment. This will be so if the system is approved with a fully decoupled payment and the requirements for maintaining set-aside are not very demanding. In contrast, farmers believe that the risk of land abandonment is rather minimal. In their opinion the key element is the maintenance of agri-environmental schemes and the implementation of other contracts in the future Nature Reserve.

Current trends do not allow us to foresee a clear future scenario. The lower productivity of the area and the high levels of fallow land seem to indicate a possible risk of land abandonment where full decoupling is introduced. Nevertheless, the available machinery, the capacities of the farmers, the possibility to increase farm size or to subcontract agricultural tasks, together with decreasing production costs, are all factors that can counteract the tendency towards land abandonment. There are also those farmers who see the reform as an opportunity to try new crops, once the payment assures income and eliminates current obligations.

Moreover, the Villafáfila Lagoons Reserve is a good example of the complex relationship between fallow and abandonment. The area covered by fallow is currently 40 per cent. Although a great number of the restrictions necessitating fallow are expected to disappear in the future, many farmers can consider decoupled payments as a sort of voluntary set-aside. The strategy related to fallow land will be a key aspect in the process of land marginalisation.

Whatever the final modality of the future agrarian policy may be, three areas must be considered in order to limit the risks of agricultural marginalisation and land abandonment:

• Farm support measures must be appropriately adapted.
• An adequate agrarian productive model, which considers other non-agrarian functions, must be promoted.
• Economic diversification must be stimulated.

Two lessons have been learnt from previous experience:

• CAP horizontal measures supported an intensive and specialised model that damaged the natural habitat.
• Agri-environmental schemes have demonstrated the ability to adapt land use to the needs of habitat conservation. A voluntary contract between farmers and public authorities seems to be a good way to support multifunctionality and incomes. However, agri-environmental measures must be improved by reducing the uncertainty of their implementation period and by avoiding competition between schemes. To overpay farmers as compensation for assuming environmental compromises and to discriminate between farmers depending on budgetary fluctuations should be avoided. In addition, agri-environmental measures could be complimented by the Nature 2000 network compensatory allowances.

Finally, economic diversification must be encouraged. The ornithological richness of the area is the main resource giving rise to the tourism potential of the area, which has been the basis for most of the diversification efforts carried out over the period 1995–2005. These efforts represent a good initial step. In this sense, the improved living standards within the area and the presence of a seasonal population have created a growing demand for a wide range of services. Agrarian multifunctionality is showing its capacity to reinforce other activities. Agriculture alone has no possibility of playing a significant role for the viability of the Villafáfila rural area.

NOTES

1. One AWU corresponds to the work performed by one person who is occupied at least 1836 hours per year.
2. The best approximation of the number of full-time farmers is make to use of the national Census, and specifically the category of farmers who devote at least 50 per cent of their time to their farm.
3. The activity rate is the ratio between active population and population older than 16.

REFERENCES

Alonso, J.C. and J.A. Alonso (1990), *Parámetros demográficos, selección de hábitat y distribución de la Avutarda (Otis tarda) en tres regiones españolas*, ICONA, Colección Técnica, Madrid.

Atance, I. and J. Barreiro (2004), *Can the CAP MTR alone assure sufficient provision of enviromental outputs from agriculture? The case of Spanish cereal steppes*, Comunication, IV Congreso de Economia Agraria, Asociación Española de Economía Agraria (AEEA), Santiago de Compostela.

Birdlife International (2000), *Threatened Birds of the World*, Barcelona and Cambridge: Lynx Edicions and Birdlife International.

De Juana, E. (2005), 'Steppe birds: a characterisation', in G. Bota, M.B. Morales, S. Mañosa and J. Camprodon (eds), *Ecology and Conservation of Steppe-land Birds*, Barcelona: Lynx Edicions, pp. 25–48.

Díaz, M., M.A. Naveso and E. Rebollo (1993), 'Respuesta de las comunidades nidificantes de aves a la intensificación agrícola en cultivos cerealistas de la Meseta Norte (Valladolid – Palencia)', *Aegypius*, 11, 1–6.

INE (2002), *Agrarian Census 1999*, Madrid; National Statistics Institute.

INE (2003), *Censo de Población 2001*, Madrid: National Statistics Institute.

INE (2004), *Padrón Municipal de Habitantes a 1 de Enero de 2003*, Madrid: National Statistics Institute.

Junta de Castilla y León (2002), *La Agricultura de Castilla y Léon*, Valladolid.

MAPA (2004), *El Modelo de desarrollo y aplicación de la Reforma de la PAC en España*, Madrid: Ministerio de Agricultura, Pesca y Alimentación.

Petersen, J.E. (1998), 'Implementing agro-environmental legislation in the European Union: an analysis of Regulation 2078/92 in Spain', PhD Dissertation, University of East Anglia.

Putnam, R. (1993), *Making Democracy Work*, Princeton, New Jersey: Princeton University Press.

Rosell, J. and L. Viladomiu (1996), *Framework for Environmentally Sustainable Rural Development in the Villafáfila Lagoons Nature Reserve (Zamora, Castilla y León)*, Barcelona: RSPB-SEO-BirdLife International.

Rosell, J. and L. Viladomiu (2005), 'Steppe birds, agriculture and agricultural policy: the case of the Villafáfila Lagoons Reserve cereal steppe', in G. Bota, M.B. Morales, S. Mañosa and J. Camprodon (eds), *Ecology and Conservation of Steppe-land Birds*, Barcelona: Lynx Edicions, pp. 283–92.

Rosell, J., L. Viladomiu and G. Francés (2000), 'Comparison of the Spanish case studies: Albacete and Zamora', in I.J. Terluin and J.H. Post (eds), *Employment Dynamics in Rural Europe*, Oxon: CABI Publishing, pp. 115–36.

Stulhofer, A. (2000), *Dynamics of Social Capital in Croatia, 1995–1999*, Washington DC: World Bank.

Viladomiu, L. (1994), 'Diez años de reforma de la PAC', *Agricultura y Sociedad*, 70, 3–20.

9. How visible is marginalisation in Europe?

Teunis van Rheenen and Floor Brouwer

The outstanding scientific discovery of the twentieth century is not television, or radio, but rather the evolution and complexity of land use. While we are very well aware that there are signs of marginalisation in the European Union (EU), so far very few attempts have been made actually to indicate in a holistic manner the extent of marginalisation. From experience we know that even the effort of defining marginalisation often leads to lively debates. This is so complex because it raises a number of fundamental questions:

- Should it be considered from a biophysical or perhaps from a socio-economic point of view?
- To what extent are these two linked?

The debate goes on at a time when the sense of urgency to come up with answers has become greater than ever before. The world food equation has changed fundamentally since the late 1990s. At no time before in the history of mankind have we faced so many challenges regarding land use at more or less the same time. Globalisation, strong trends towards urbanisation, the demand of agricultural products for fuel production, high economic growth in many parts of Asia and Africa resulting in shifting diets, and climate change are some of the major global forces of change effecting land use in the rural areas in the developing and developed world.

Also, in the rural areas of Europe, depopulation has been a trend for a long time and this has taken on alarming proportions, especially considering the multifunctional role that rural areas have. Not only agricultural production, but also maintaining biodiversity, landscape and recreation, just to mention a few, are all important functions of the rural areas. Both population density and the decline of population in the rural areas to a large extent determine the ability of rural areas to fulfil the functions which society expects it to.

Agricultural policies which for such a long time have contributed to keeping farmers in the rural areas of Europe have increasingly come under attack. Increasingly, there has been a shift away from only considering the rural areas as a place where agricultural production takes place to the recognition that the rural areas play many other roles. This has led to the introduction and acceptance of the term 'multifunctional land use', but translating this concept into practical actions is not always an easy thing to do. While several studies (for example OECD 2001, 2003) have attempted to translate this concept into practical relevance for policy-makers, it seems fair to say that a lot still needs to be done in this area. Simply attempting to answer the question: 'Where in Europe can we expect marginalisation to occur in the coming 20 years?' often does not go beyond the realms of theoretical and intellectual debates and in the end it goes unanswered. Consequently, defining proper policies that will address the problem in a preventive manner also becomes difficult, if not impossible.

To move forward – albeit recognising the potential shortcomings of quantitative assessments – this chapter uses an indicator approach to identify areas which are at higher risk than others to becoming marginalised during the period 2010–20. To achieve this we have selected four indicators which we consider relevant for the future marginalisation of rural areas in general and agriculture in particular. Subsequently, we identify those areas that pass a threshold value for these indicators, and could potentially be at risk of becoming marginalised from an agricultural perspective.

RISK OF MARGINALISATION: IDENTIFICATION OF ACTUAL HOTSPOTS

Regions can be identified that are more vulnerable to the occurrence of marginalisation than other areas. This chapter focuses mainly on the 'initial determination' of actual hotspots. Following an initial identification of hotspots, further analysis would be required to determine to what extent the perceived problem actually needs to be addressed. So in a way, the approach followed in this chapter could be seen as an early warning system towards marginalisation.

Chapters 5 to 8 of this volume clearly show that the relevant indicators differ across countries. While in many cases sufficient data may be available to describe the indicators, often it remains difficult to determine the threshold values, that is, those values that will clearly indicate when there is reason to assume that there is marginalisation of land. Basically, to give a good estimate of how much land will be at risk of becoming marginalised, we need to have information on the following three core questions: 'How likely?' 'How severe?' and 'When?'

There is no shortage of elaborated descriptions of different indicators that each in their own way attempt to capture insights regarding marginalisation; however, only a very few studies have actually gone beyond this stage and really provided evidence concerning the extent to which marginalisation of agriculture or of the rural areas in general can be expected. Two exceptions are Pyykkönen (2001) and Pinto-Correia et al. (2006) who, albeit with different indicators, developed threshold values and techniques of analysis which were used to quantify marginalisation. The former concluded that up to 10 per cent of the agricultural land in Finland is marginalised and the latter concluded that up to a quarter of the Portuguese continental territory is not only considered marginalised from an agricultural perspective but also from a rural development perspective in general.

This chapter addresses marginalisation with a current and forward-looking perspective, that is, the identification of rural areas which we think, based on preliminary analysis, are more highly at risk of becoming marginalised than others. Baldock et al. (1996) made an effort to give an indication of those regions in Europe that are susceptible to marginalisation. They used cluster analysis and the following five broad sets of indicators to determine these regions:

- biophysical conditions;
- agricultural land utilisation;
- farm structure;
- farm income; and
- rural and regional development.

Their conclusion was that two types of regions are considered susceptible to marginalisation; namely one type which is mainly characterised by extensive agriculture, and the other which mainly includes small-scale farming. The methodology proposed in the chapter includes the quantification of a so-called Integrated Marginalisation Risk Indicator (IMRI) which – with a high level of level of abstraction – will indicate areas at risk of becoming marginalised during the period 2010–2020. The potential hotspot identification would need to be followed by more location-specific studies. The IMRI uses a combination of the following indicators that all are based on regional averages:

- Agricultural income as a percentage of non-agricultural income. Low income in agriculture and in rural areas is considered as being an important indication of marginalisation. Vulnerability of agriculture to marginalisation may increase in cases where farm income is substantially below income in the rest of the economy.
- Share of agricultural landholders that are older than 55 years. Age of a farmer and having a successor are crucial for continuing agriculture.
- Population density expressed as inhabitants per square kilometre. Population density is a crucial indicator for the availability of public and private services to its inhabitants.

- Erosion expressed as tonnes per hectare per year. Erosion is an important indicator to consider in many parts of the EU. Soil erosion is a core issue for soil destruction and loss of land for agricultural production. Farm support programmes are implemented through the Common Agricultural Policy (CAP) to compensate farmers for measures they take to combat soil erosion. Particularly for the new accession countries, it will be interesting to see how and when new driving forces for land use come into play and how they will influence marginalisation indicators such as erosion.

These indicators are considered relevant as driving forces in the marginalisation processes, and by combining them we can identify those areas where we expect problems to arise, as shown in Figures 9.1 to 9.4.

A more detailed picture regarding the indicators is presented at the national level. Based on regional averages in the EU-15, each of the four indicators has a share of between 20 and 30 per cent (Table 9.1). In the EU-15, some 30 per cent of total population lives in areas with a population density below 50 inhabitants per km^2. Also, farm income from about 30 per cent of the farmers in the EU-15 is less than half of the average income in the rest of the economy, and soil erosion exceeds 2 tonnes per hectare on around a quarter of the territory.

Table 9.1   Distribution across countries of four indicators with risk of marginalisation

Country	Share of population below 50 inhabitants per km2 (%)	Share of regional average age of farmers aged over 55 (%)	Share of farmers with income less than half of income in rest of economy	Share of land with erosion over 2 t/ha/year
Austria	0.0	0.0	56.7	0.0
Belgium	0.0	0.0	11.2	11.2
Denmark	0.0	0.0	4.2	78.4
Finland	90.4	0.0	9.0	n.a.
France	28.8	0.3	10.3	15.9
Germany	0.0	0.0	37.3	5.1
Greece	53.3	47.0	82.4	60.1
Ireland	90.3	0.0	56.4	0.0
Italy	4.4	91.2	34.1	49.0
Luxembourg	0.0	0.0	100.0	0.0
Netherlands	0.0	0.0		0.0
Portugal	67.2	96.4	39.5	82.7
Spain	66.4	24.8	28.0	46.2
Sweden	57.0	0.0	29.2	n.a.
United Kingdom	0.0	5.2	24.0	0.0
EU-15	30.2	19.5	27.8	24.7

Note: n.a. – not available

Source: Own calculations, based on FADN; Eurostat, Luxembourg.

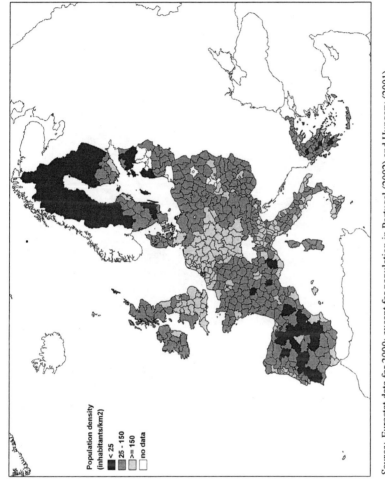

Source: Eurostat data for 2000; except for population in Portugal (2002) and Hungary (2001).

Figure 9.1   Population density (inhabitants per km²)

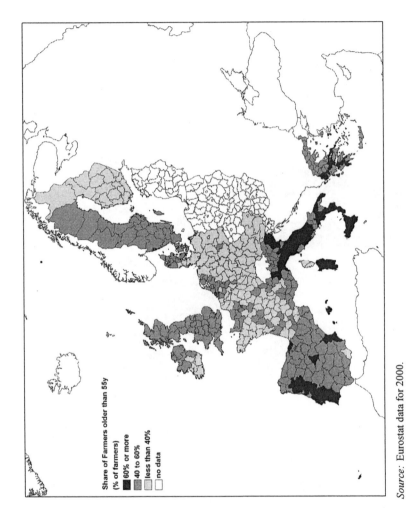

Figure 9.2 Share of holders aged 55 years and over (%)

Share of Farmers older than 55y
(% of farmers)

■ 60% or more
▨ 40 to 60%
▢ less than 40%
□ no data

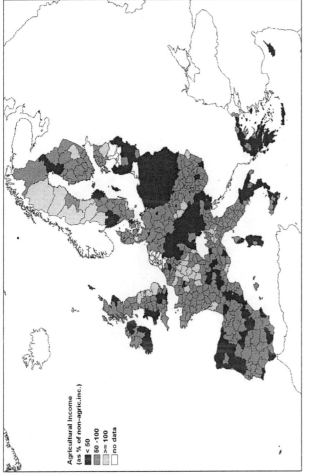

Note: = *gross value added per worker in agriculture / gross value added per worker in non-agriculture (industries and services).*

Source: Eurostat data for 2000; except for employment in Greece (2001), Malta (2001) and Poland (2002) and for gross value added in Malta (2001).

Figure 9.3 Agricultural income as percentage of non-agricultural income

177

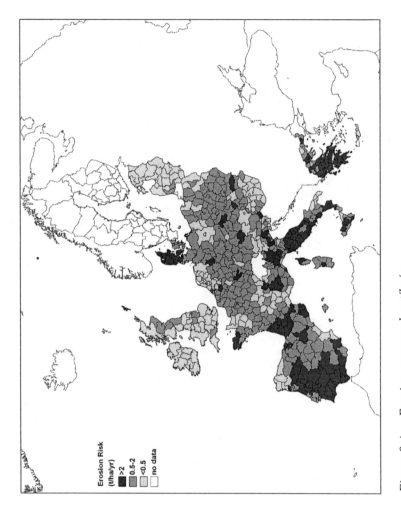

Erosion Risk
(t/ha/yr)
■ >2
▨ 0.5-2
▨ <0.5
□ no data

Figure 9.4   Erosion expressed as t/ha/yr

Given current agricultural markets, socio-economic and environmental conditions, Table 9.2 indicates that there is no risk to marginalisation in regions covering almost two-thirds of the land in use in the EU-15. Risks of marginalisation are high on around 5 per cent of the land, which is mainly due to unfavourable conditions in Portugal, and to a lesser extent also in Greece and Italy. Issues at stake are climate change (shortage of precipitation is a major issue in the Mediterranean part of Europe) but also trade and agricultural policies in relation to emerging economies and global stocks.

Table 9.2   Size of territory vulnerable to the risk of marginalisation (low, medium and high risk) (share of national total, in %)

Country	Low	Medium	High
Austria	56.7	0.0	0.0
Belgium	0.0	11.2	0.0
Denmark	82.6	0.0	0.0
Finland	81.4	9.0	0.0
France	34.2	10.6	0.0
Germany	36.1	3.2	0.0
Greece	27.6	23.1	23.7
Ireland	53.3	46.7	0.0
Italy	27.8	59.9	10.4
Luxembourg	100.0	0.0	0.0
Netherlands	0.0	0.0	0.0
Portugal	13.8	3.9	64.9
Spain	36.6	55.6	5.9
Sweden	77.7	0.0	0.0
United Kingdom	29.2	0.0	0.0
EU-15	36.2	23.1	4.9

Note: Low – at least one factor is critical.
Medium – at least two factors are critical.
High – at least three factors are critical.

Source: Own calculations, based on FADN; Eurostat, Luxembourg.

CONCLUSIONS

The many dimensions of the marginalisation of land, which are often driven by global forces, yet have area-specific characteristics, make it a challenging topic to address. Yet we can not shy away from these challenges because too much is at stake, and also prevention is better than damage control. Approaching the marginalisation risk of land in an integrated indicator fashion is not that common, yet we think that this is a promising area for further research.

When considering the rural areas in the EU, and indeed in many other areas of the world, we generally now accept the fact that they are multifunctional. The concept has become part and parcel of policies and debates that concern rural areas and in many countries, such as Austria, it has become the core of agricultural and rural policies.

Some areas may – at present – not really be affected by land abandonment and depopulation; however, in recent years their vulnerability has increased and this has occurred in ways that at first sight are not all that obvious. For example, almost all of Austrian farmers are eligible to receive compensation. A high 'compensation dependency' goes hand in hand with a high vulnerability to marginalisation. If support measures were to decrease, this would potentially increase the risk of land abandonment and depopulation.

In several countries, especially in the new EU member countries, marginalisation is only recently beginning to receive political or indeed scientific attention. In some of these countries, the political developments have been so dramatic that analysis of trends in indicators is complicated. Often data are still insufficiently available, and to get an indication of the extent to which marginalisation is actually a problem, criteria as developed for Less Favoured Areas (LFA) are at this stage probably the most appropriate. For example, in the Czech Republic, more than half of the land can be considered as marginalised on the basis of LFA criteria. In a number of new member countries, such as the Czech Republic, at the moment there is a lack of proper instruments and measures to combat marginalisation. In a country such as Estonia, recent and future anticipated changes in the rural countryside should be carefully monitored. The changes have been dramatic with far-reaching effects on landscape and biodiversity. The drastic decrease in employment in the agricultural sector will have far-reaching consequences for the rural countryside in general. However, it is not only in the new member countries that we expect big changes. As already mentioned, studies show that marginalisation has already progressed substantially in several other European countries such as Finland and Portugal, and our analysis shows that many other areas of the Mediterranean are very much at risk of further marginalisation. The Portuguese study shows that marginalisation is a process which is not only dependent on the agricultural sector. For different countries, different criteria are used to determine to what extent land is actually marginalised. For example in Spain when the number of inhabitants per square kilometre is lower than ten, a situation has developed that is not considered sustainable.

REFERENCES

Baldock, D., G. Beaufoy, F. Brouwer and F. Godeschalk (1996), *Farming at the Margins: Abandonment or Redeployment of Agricultural Land in Europe*, London: Institute for European Environmental Policy, The Hague: LEI-DLO.

OECD (2001), *Multifunctionality: Towards an Analytical Framework*, Paris: Organisation for Economic Co-operation and Development.

OECD (2003), *Multifunctionality: Policy implications*, Paris: Organisation for Economic Co-operation and Development.

Pinto-Correia, T., B. Breman, V. Jorge and M. Dneboska (2006), *Estudo sobre o Abandono em Portugal Continental, Análise das dinâmicas de ocupação do solo, sector agrícola e comunidade rural, Tipologia de Áreas Rurais*, Évora: Universidade de Évora.

Pyykkönen, P. (2001), 'Maatalouden rakennemuutos eri alueilla' (Regional differences in structural change of Finnish agriculture), Pellervo Economic Research Institute Reports No. 180, Helsinki, Finland.

PART III

SUSTAINABLE LAND MANAGEMENT
PRACTICES

10. High nature value farming and the agri-food chain in Japan

Yoichi Matsuki and Miki Nagamatsu

Japan is a long, narrow chain of islands stretching 3300 km from the subarctic to the subtropical zones. Mountainous areas account for 67 per cent of the total territory of the country. A limited part of the country is flat, and this is also highly urbanised. Farmland covers about 13 per cent of the total territory of the country, and the cultivated land per farm household is very small (on average about 1.8 hectares). The small size of farming operations and difficulties in controlling water use on the individual holdings required collective control and use of water, facilitating the formation of farming communities. Community rules established to ensure smooth operations had a considerable influence in fostering the spirit of mutual aid and creating and passing on traditional rural cultures. The need for reorganising regional agriculture has arisen in Japan because of overproduction of rice and other farm products, and of urbanisation in rural areas since the late 1960s. Production adjustment measures have led to the abandonment of cultivated land, such as paddy fields, upland fields, temporary meadows and land under permanent crops. By the year 2005, abandonment of cultivated land amounted to 384 700 hectares. Since the mid-1990s, land abandonment had increased by 140 700 hectares (57 per cent). Abandoned cultivated land is currently almost 10 per cent of total cultivated land. Ageing of the rural population increased more rapidly than that of the urban population because young people tend to leave rural areas to live in urban areas. Problems arising from the increased share of elderly people are more prominent in rural areas than in urban areas. In short, the regional agricultural structure has been changed in four ways:

- collapse of the cooperation system of family farming and agricultural community at the hamlet level;
- transformation of farmland to urban land;
- increase of the non-farming population in rural villages; and
- reduction of nature in farming and the related decline of multifunctionality.

The self-sufficiency rate of food in Japan has sharply declined during the past 40 years. Since the mid-1960s, this ratio went down from 73 per cent to 40 per cent in 2007 (when measured in the supply of calories). On a grain basis it reduced from 62 per cent to 28 per cent. From a long-term standpoint, one of the major factors behind this declining trend is the fundamental change taking place in Japanese dietary patterns, as reflected in the increasing consumption of animal products and fats and oils. These products are largely dependent on the import of feed grains and oilseeds, because of the limited land resource for food production. Rice consumption shows a decreasing trend over time. Farm structure, and especially the lack of farm labour and the ageing of farm population, have also caused a decline in the supply of domestic farm products. To secure a stable food supply, it is essential to make efforts to increase domestic production. Not only farmers, but also consumers, manufacturers, distributors and other related parties should create new agri-food chains together which produce much better quality food than imported food.

This chapter deals with current developments of the agri-food system by several different chain actors to support domestic agriculture and rural society. We will review current trends in the agri-food system of Japan. One of the interesting aspects is the interplay between agri-products and nature conservation. Such trends are put in the context of recent changes in food production both nationally and internationally. There are national policies, strategies and plans that have facilitated the different approaches. Focus is applied to the different actors supporting domestic agriculture and rural society under the New Basic Law and in response to trade requirements from the Uruguay Round Agreement on Agriculture (URAA). The Basic Law on Food, Agriculture and Rural Areas was introduced in 1999 following reviews of the post-war agricultural policies under the Agricultural Basic Law. This new law introduced a policy scheme that is based on four new principles:

- securing a stable food supply;
- fulfilment of multifunctionality of agriculture;
- sustainable development of agriculture; and
- promotion of rural areas.

Japan signed the URAA, and the principles of the newly enacted Basic Law on Food, Agriculture and Rural Areas and accompanying measures must properly comply with the rules on global trade. The Japanese government strongly supports multifunctionality of agriculture, including features such as food security, environmental protection, viability of rural areas, and food safety.

In order to revitalise agriculture and rural society, national and local governments, farmers, consumers, manufacturers, wholesalers, retailers and any other related parties should work together to develop new economic systems like direct agri-food supply chains in partnership with producers. See also Baldock et al. (1996), exploring strategies to cope with the marginalisation of agricultural land, and Brouwer and Lowe (1998), exploring the broader European context of agricultural policy reform and the environment.

MANAGEMENT CONCEPTS PRESERVING BIODIVERSITY

High Nature Value Supported by Farming

Agricultural production has long been an economic activity intended to nurture the life of living things, not to manufacture goods artificially as in industry. Although individual species can only survive in habitats under diverse bio-ecological conditions, human beings are able to control the growing conditions of life outside their original ecological systems, meanwhile contributing to the pollution of the environment and the reduction or even extinction of species. This in turn also harms human life and public health. Such a deviation from the original character of farming has stemmed from the adoption of practices which exterminate and extinguish diverse forms of life, in applying insecticides and chemical fertilizers so as to exclusively grow a specific living thing without proper perception of ecological principles. Baldock and Beaufoy (1993) offer an review of new perspectives on the linkages between nature conservation and European agricultural policy. Only farming following the original ecological principles is sustainable, so as to 'utilize the human, natural and artificial resources to satisfy the needs of current generation without jeopardizing the productive capacity for future generations'. We regard this sustainable farming as the basic conception of agricultural production technology to conserve biodiversity on the farm.

The Concept of Nature Management Farming

The concept of nature management farming can be regarded as a rather new movement. It started in the Netherlands and the United Kingdom from the 1980s and through the mid-1990s. Its basic concept is that farmers, who manage the land, should also be the main contributors of the management of nature, instead of nature protection organisations. Nature management farming can be defined as land management practices that support multifunctionality of the land. Farming provides food and fibre, maintains agricultural landscapes, generates employment in rural areas, supports the bio-ecological system and biodiversity, and also controls the quality of water, air and soil, and animal welfare. Farming has the multifunctional nature to generate these products and services in an integrated manner.

Multifunctionality of agriculture is emphasised all over the world and not limited to Japan and European countries. As part of the World Trade Organization (WTO) negotiations there are signs that payments for public services are being allowed to be given to promote multifunctionality. Movements are being initiated to re-evaluate functions like the conservation of biodiversity. Multifunctionality can be divided into five main functions:

- supply of safe and secure (anxiety-free) agricultural products in a stable manner;
- ecological function coexisting with biodiversity values;
- supply of educational material and the provision of information for scientific research;
- cultural and landscape aspects (which is less debated in Japan than in Europe); and
- outdoor recreation function.

Nature management farming is not limited to the first function (that is, supply of agricultural products), but embraces all the other functions as well.

Preservation of Biodiversity by Farmers in Practice

The projects on developing nature management farming will support a farming system, with farmers carrying out activities preserving biodiversity among various productive as well as service-providing functions. Its objective is to promote multifunctionality in practice through the actual operation of developing it as part of management and operation activities. It also aims to protect the environment through the basic objective of agricultural policies, including enhanced public functions and harmonisation with the environment. Such farming systems provide agricultural products and agri-environmental service goods with high nature value in an integrated manner while preserving the local natural ecosystems.

A Master Plan to Support Nature Management Farming

In addition to the need to supply food, agriculture also needs to adopt farming practices respecting the environment (for example avoiding pollution from the use of insecticides, chemical fertilizers and disposal of animal waste). Moreover, citizens demand that farmers prevent the extinction of wild animals and plants, preserve biodiversity and strengthen landscape preservation, recreation and the provision of residential land.

- The basic concept of nature management farming in support of agriculture and mountainous rural areas is defined as follows:
  - (a) Utilise specific local natural resources (for example land, water, climate, animals and plants), and to provide primary products from agriculture and forestry with high added value and high nature quality.
  - (b) Establish high standards of living in rural and mountain areas for the local population through the proper management of natural resources. This includes the supply of environmental goods and services (for example natural landscape, habitats of wild animals and plants, historical and cultural landscape, recreation sites and so on) to local visitors.
  - (c) Provide consumers with a living environment (air, water and energy) which is generated by regional natures and to develop and provide new functions thus created to urban societies.
- Adopt the principles of organic farming that are compatible with local natural ecological systems. Moreover, livestock husbandry should consider animal welfare requirements, securing their physical and mental health.
- Farms should develop management plans, taking into account land utilisation and nature management farming. This implies that agricultural and forest lands preserve natural ecological systems. Income losses from nature management farming should be compensated for. The income compensation method would be carried out through contracts and agreements among farmers, consumers (reflected in food prices) or taxpayers (that is, subsidies from state budgets or local governments) or fund contributors (that is, trust funds and so on).
- Establish integrated land utilisation plans at the local level for housing, agriculture, forestry and nature protection. A network of nature management farms is built among biotopes (that is, natural protection zones).
- Pluriactivity of rural households is achieved and rural employment is expanded.

On the basis of the citizens' aspirations for multifunctionality of agriculture and forestry and rural and mountain areas in Japan and the concerned policy response the achievement of high nature farming is considered to be very likely, but there are four main challenges:

- Develop a land utilisation plan for high nature value agriculture and forestry that integrates high nature value farming with other activities (including industry).
- Identify the management bodies that could act as spearheads of the concept of high nature value farming.

- Formulate the capital investment for multiple management of the high nature value farms, involving agriculture, forestry but also other various industries.
- Identify marketing strategies to sell the various products and services of the high nature value farms.

PARTNERSHIP BETWEEN ORGANIC FARMERS AND CONSUMERS

Three cases are presented relating to partnership between the agri-food supply chain and consumers (see also Nagamatsu and Matsuki, 2003 for an elaborated discussion on cooperative approaches in Japan).

An Agri-food Chain Driven by Consumer Cooperatives

The cooperatives have jointly developed a membership system of five to ten households who place advance orders for food. Japanese consumers seek agricultural products and processed food of high quality. Through these businesses, they have developed new concepts regarding the high quality. A movement between consumers and agriculturalists has started to create new concepts of nature symbiotic agriculture and high nature value.

Safety of Agricultural Products and the Development of Organic Farming

The organic farming movement started in Japan during a period of high economic growth. Environmental pollution and contaminated food were observed as a negative aspect of the successful economic growth. In particular residual chemicals in drinking milk made consumers feel concerned about food safety. Nowadays, a similar situation has arisen in Japan with the debate about genetically modified food and the use of hormones. Farmers also face health problems from the use of agricultural chemicals. In looking for safe foods as a result, consumers were searching for organic producers, and the organic farming movement started. The movement toward organic farming gradually expanded throughout the country and the Japan Organic Farming Research Group was established in October 1971. Members of this group selected a marketing pattern directly connecting producers to consumers, rather than the ordinary marketing pattern through wholesale markets. This pattern can be regarded as the prototype of the 'producer–consumer coalition'. Such a coalition can be built on trust between producers and consumers. The coalition is established on ten principles:

- friendly and mutually helpful relationships between producers and consumers;
- production plans are developed jointly with consumers;
- all food supplies are purchased by consumer groups; among others through the organisation of meals;
- the consumer groups guarantee prices to producers;
- mutual understanding between producers and consumers;
- transportation of products is organised by the producers themselves;
- management of producers' and consumers' groups based on democratic principles;
- emphasis on learning activities in producers' and consumers' groups;
- groups are maintained at a reasonable size; and
- producers and consumers jointly aim for the achievement of their ideals.

Spiritual aspects are emphasised more than technological aspects. Organic farming thus becomes a movement seeking for a philosophy and a view of the world beyond an issue of farming method.

A Coalition of Enterprises from the Producer to the Consumer

Women play a key role in consumer activities. They receive the shipped products that come directly from the producers and they attend courses. These activities are rarely carried out by men, who work during daytime. A lot of effort is involved, and people need to be very committed to continue in the long run. The increasing participation of women in the labour market implies that they cannot afford to be involved in activities related to food alone. Accordingly some people want to buy organic products in more convenient ways rather than through the producer–consumer coalition. Concerns with the safety of food have been enhanced by the events such as the contamination of food and the environment by radioactive fallout from the accident at the Chernobyl nuclear generation plant, and the application of pesticides after harvesting to imported agricultural products. The Association Protecting Earth, which is one of the leading organisations of the organic farming movement, has developed the movement to include enterprises by establishing a limited company, Daichi (Great Earth), in 1978. These enterprises diversified their operations during the 1980s, including by wholesaling products to stores specialising in natural food and some supermarkets, and through the supply of food for school lunches. Moreover as the number of people unable to participate in the joint purchase of organic products increased, they were organised into individual delivery systems using home delivery. As it becomes clear that the producer–consumer coalition could be transformed to enterprises, traders entered into this market as they could sell vegetables at high prices. Before the standard of organic products was established in the Japan Agricultural Standard (JAS) Law, sometimes certifying seals of organic products were sold on wholesale markets and put onto products produced under conventional cultivation, which were sold in retail stores. At present, marketing channels of organic agricultural products are diversified to

- producer–consumer coalitions;
- specialised stores for natural food;
- consumers' cooperatives; and
- supermarkets and department stores.

In the food sections in department stores, imported organic food faces strong competition with that produced domestically.

Since the coalition between producers and consumers was the main stream of organic production, consensus was reached on the basis of mutual understanding. In a case where pesticides had to be applied, an inspection and certification by a third party was not required. However, as the guidelines of organically produced products were agreed by the Codex Alimentarius Commission, inspection on compliance was needed and a certification system of organically produced food was introduced. It actually began to be implemented in 2001 when the Japan Agricultural Standard (JAS) Law was amended.

HIGH NATURE VALUE FARMING AND RICE PRODUCTION

Ibaraki prefecture, which is some 100 km from Tokyo, has a winter migration site for oohishikui (bean geese) in the Kanto area. The place is significant as it represents the southern limit of winter migration sites of these geese on the Pacific coast. A small flock of 50 to 70 birds currently fly there every winter, and it is generally said that the flock of geese risk extinction when a flock falls below 1000 birds. Therefore, the group in Edosaki town is at the brink of extinction. In order for numbers to recover, a large area would be needed for the birds to pass the winter period, including a diversified water area environment with feeding places, roosts and refuges.

The market price of rice grown in this area includes a premium for the protection of the geese, which is around 20 to 25 per cent. Order forms for the rice are sent to consumers in December. The rice is grown on land where ploughing is not carried out during the winter period when the geese are passing, and the rice is produced in this area with reduced application of pesticides. Husked rice is bought from the cooperating farmers at a price that includes compensatory payments. Table 10.1 shows the dynamics of the shipped amounts and the number of buyers. The rapid increase between 1997 and 1998 was largely achieved by the wide dissemination of information about this type of rice through extensive reporting in the media.

Table 10.1 Amount of rice transported and number of buyers

Year	Number of orders	Shipped amount (kg)	Approximate number of buyers
1997	550	2750	160
1998	1554	7770	560
2004	1500	7500	300

In order to protect the geese in this region and to strengthen the viability of rice production further, the following issues should be addressed:

- designate part of this region as a protection site for wild birds and animals;
- take measures that prevent non-cultivated paddy fields to increase habitat functionality for biodiversity;
- reduce variation in the quality of rice between farmers and over time;
- enhance the awareness of farmers and strengthen the efforts to improve product quality (taste and organic farming);
- secure successors of farm households;
- advance the announcement date of allocation of this type of rice to farmers;
- review and strengthen the system of buying, packing and ordering.

These tasks are currently carried out on a voluntary basis, but a further increase in the workload would become too demanding for volunteers. Organisational and marketing methods should be reviewed and strengthened in future as the number of purchasers increases.

CONCLUSIONS

Since the 1990s, the concerns of consumers in Japan have moved from food safety towards the protection of nature. They are now not only interested in agricultural products not depending on pesticides and chemical fertilizers, but are also concerned to strengthen agriculture and rural areas that protect nature and biological diversities against from being contaminated by modern farming methods. Cooperation between consumers and farmers has increased, and the recognition of high nature value farming has increased in Japan.

Cooperatives in this country have benefited from the development of consumer-driven agri-food supply systems since the 1960s. Nowadays, they try to revitalise rural communities by making new agri-food supply chains which can produce environmentally friendly food. Major trading companies which are engaged in importing food are starting to adopt new strategies of promoting domestic agri-food chain development. Especially, they are interested in building an ecological and recycling society. The recycling system must be realised by integrating the agri-food supply chain and a good waste management chain. The food industry is also required to make efforts in reducing damage to the environment caused by its operations, through waste reduction and recycling. A certain mechanism needs to be developed so that the food industry, farmers, consumers and national and local governments can work together in the socio-economic system founded on a cyclical use of resources by role sharing.

In order to recover an agriculture which is friendly to animals and preserves biodiversity, and at the same time responds to the needs of modern society, the following basic principles should be established:

- farm animal welfare should be considered in terms of five freedoms, which are freedom from hunger and thirst, freedom from discomfort, freedom from pain, injury or disease, freedom to express normal behaviour and freedom from fear and distress; and
- wild plants and animals are exposed to the risk of extermination since their habitats are invaded by the economic activities of human beings.

In addition to the above principles, the following goals could be set:

- achievement of nature management farming, in which farming coexists with wild plants and animals;
- collaborative development of livestock products considering animal welfare, and high value-added food chains though nature management farming; and
- achievement of direct payment for nature management farming and livestock husbandry which considers animal welfare.

This all needs to contribute to strengthen farming practices that coexist with biodiversity in rural areas, to improve animal welfare conditions, preserve biodiversity and respond to new societal demands.

REFERENCES

Baldock, D. and G. Beaufoy (1993), *Nature Conservation and New Directions in the EC Common Agricultural Policy*, London and Arnhem: Institute for European Environmental Policy.

Baldock, D., G. Beaufoy, F. Brouwer and F. Godeschalk (1996), *Farming at the Margins: Abandonment or Redeployment of Agricultural Land in Europe*, London: Institute for European Environmental Policy; The Hague: LEI-DLO.

Brouwer, F. and P. Lowe (1998), 'CAP reform and the environment', in F. Brouwer and P. Lowe (eds), *CAP and the Rural Environment in Transition: A Panorama of National Perspectives*, Wageningen: Wageningen Pers, pp. 13–38.

Nagamatsu, Miki and Yoichi Matsuki (2003), 'Food safety and security system in agrifood chains in Japan', in A.G.J. Velthuis, L.J. Unnevehr, H. Hogeveen and R.B.M. Hiurne (eds), *New Approaches in Food-Safety Economics*, Dordrecht: Kluwer Academic Publishers, pp. 129–32.

11. Social capital: a dynamic force against marginalisation?

Georg Wiesinger, Hilkka Vihinen, Marja-Liisa Tapio-Biström and Gábor Szabó

During the last 20–30 years in many European countries, rural and agricultural marginalisation has accelerated due to structural changes, political processes and socio-economic development. Rural marginalisation is a major problem in many less-favoured and remote European regions. The closing down of farm enterprises, depopulation, land abandonment and land use change have economic, socio-cultural and environmental implications. For a great part there is an unanimous understanding that the process of marginalisation depends on socio-economic development, regional, agricultural and environmental policies, but also on global trends and on so-called intrinsic factors of local communities which refer to the importance of local social capital and networks.

Marginalisation can be largely explained by unfavourable conditions and missing resources, but not entirely, and not in all regions. Some regions with a very sparse population, lack of policy support, poor economic conditions and unfavourable climatic conditions prove to be more viable than regions with much better circumstances. Proper policy measures and environmental qualifications may to some extent explain why the situation is not as bad as some of the indicators would suggest. But social capital in terms of social networks and generalised norms of reciprocity (Putnam, 2000) might also play a crucial role.

This chapter evaluates whether the social capital approach could be a means to fill the missing link and to gain a better understanding of marginalisation by connecting policy measures and social capital with marginalisation. The conclusions are mainly drawn from the experience of three case studies on strategies to cope with marginalisation, which have been conducted in Austria, Finland and Hungary.

CONCEPTUALISING SOCIAL AND POLITICAL CAPITAL IN RURAL DEVELOPMENT

Scientific discussion on social capital (Bourdieu, 1979, 1986; Coleman, 1988; Putnam, 1993, 2000; Woolcock, 1998; Fukuyama, 1999; Burt, 2000; Norris and Inglehart, 2003; Árnason et al., 2004; Tillberg-Mattsson and Stenbacka, 2004) has emphasised the importance of social ties and shared norms to societal well-being and economic efficiency, and the concept has been widely used in the study of social inequality and hierarchical social structures. Putnam (1993, 2000) above all linked the idea of social capital to the importance of civic associations and voluntary organisations, and emphasised positive aspects of social control.

According to Bourdieu's theory of capital (1986), economic capital is immediately and directly convertible into money and may be institutionalised in the forms of property rights, while cultural (or intellectual) capital may be institutionalised in the forms of educational qualifications. Social capital is made up of social obligations and connections. Social capital may be institutionalised in the forms of a title of nobility and can be defined as the aggregate of the actual or potential resources which are linked to possession of a durable network of more or less institutionalised relationships of mutual acquaintance and recognition. Social capital, like other types of capital, is unevenly distributed, mobilised, utilised, transformed and exchanged in society.

Recently Árnason et al. (2004) discussed the concept of social capital in the context of rural development. They consider this concept as an attempt to capture the non-economic aspects of society that promote economic growth or more widely positive effects. Social capital may affect the performance, competitiveness and social cohesion of a community. Networks can be understood as articulating the flows of information and resources that produce rural development and society more generally. Focusing on networks therefore will allow investigation of the mechanisms by which people capture or contain benefits of development, and also illustrates the ways in which people in rural areas are becoming linked in complex ways to different scales of economy and society.

The intangible asset of social capital can be affected by policy, both positively and negatively. Measures which encourage the creation of networks and working modes enhancing cooperation are important elements in the creation of social capital. On the other hand policies which encourage competition and divide rural inhabitants into winners and losers may be detrimental to the positive development dynamics and to the rural social fabric. There are some examples of this from United States (US) agricultural areas (Schubert, 2005).

Different criteria of citizen participation to public processes are worth noting since they constitute elements in favour of the dynamism and the institutional innovation of local society. The number of associations per number of local inhabitants was suggested by Putnam (1993) as a benchmark for social capital. Yet, this tells very little about the structure of the local population which is attracted to associations or about the gender issue. Some inhabitants can be members in several associations or local networks, while others are socially excluded. Women, young people and persons with special lifestyle or cultural interests may find themselves not affected by the given associations. On the other hand women are often the promoters of social cohesion within both their formal and informal networks. In the LEADER+ programme it has been noted that, for example in Finland, women dominate the local action groups (LAGs) and the various development activities supported by the programme (LEADER+ mid-term review evaluation).

Norris and Inglehart (2003) emphasised that associational membership can be vertically and horizontally segmented for women and men. Yet, there is the threat of patron–client politics, personalism and centralisation of decision-making power in contexts lacking social capital. In the absence of horizontal solidarity, vertical dependence appears to become a rational strategy for survival. This implies that there is also a qualitative aspect in social capital. As much as local communities are prone to embrace their citizens and assist persons in need, their close ties also account for dynamics of social exclusion of all those who are unable or unwilling to cope with the strict norms of the community. Another potential dark side of social capital is that local communities are hostile towards incomers and thus prevent innovations.

Political scientists have criticised Putnam (for the discussion, see for example Booth and Richard, 1997) for paying too little attention to how the socio-political context may shape civil society and for having the causal sequence backwards – rather than civil society shaping government performance, the state has the capacity to stimulate high levels of organisational activity. To a large extent, political capital has been proposed as a means to overcome some of the problems of using social capital as a catch-all concept for explaining the importance of non-material factors in development. Like social capital, political capital is also acquired through participation in voluntary associations, but only when these organisations also engage in political activity, when there are defined goals and when associational activism fosters attitudes and behaviours that actually influence political regimes. For Booth and Richard (1997), political capital indicates such state-impinging attitudes and activities.

Although political capital has not apparently gained as wide currency in political science as social capital in sociology, it is increasingly recognised as one potential remedy to the limited use of political analysis in studies of development. It may highlight, together with social capital, the missing dimension of the marginalisation process as well. The most extensive elaboration of political capital comes from Birner and Wittmer (2003), who make a distinction between 'instrumental' and 'structural' political capital. Instrumental political capital consists of the resources which an actor can dispose of and use to influence policy formation processes and realise outcomes which are in an actor's perceived interest. Structural political capital refers to the structural variables of the political system which influence the possibilities of diverse actors to accumulate instrumental political capital and condition the effectiveness of different types of political capital.

Rakodi (1999) defines political capital as 'based on access to decision-making', because of the significance attached to powerlessness in the poor's own definitions of poverty. To affect one's lot, a person or a group has to have a voice to raise an issue and to articulate arguments, and needs to have access to an arena where decisions are taken. If people feel empowered as regards these acts and arenas, even in as difficult a situation as rural and agricultural marginalisation, the prognosis for their case is more positive. Baumann (2000) states that the notion of political capital is critical in linking structures and processes to a local level. According to Baumann, political capital links an individual or a group to power structures and policy outside the locality, and explains where local people are situated in terms of the balance of power in relation to other groups.

CASE STUDIES

Some of our considerations will be exemplified in the following by case studies in Austria, Finland and Hungary.

Neunkirchen: A Case Study in Austria

The Neunkirchen district is located in the province of Lower Austria at the eastern outskirts of the Alps. At first glance this region does not appear to be extremely marginalised when observing economic, topographic or physical conditions. There are no very high mountains, steep slopes or frequent natural hazards. The accessibility of the region is quite favourable; most villages are rather close to Vienna (only 60–90 km) so people can commute.

But there is also a reverse side. Only 31 per cent of the total cadastre area of about 1150 km² is suitable for permanent settlement. The situation in the region is very heterogeneous. On the whole the number of inhabitants (about 86 000) is quite stable. But population increase in the hilly and flatland areas in the north and north-east is contrasted by a quite strong decline in some inner-mountain villages in the south and south-west. The proportion of forests and woodlands is extremely high, particularly in the mountain areas of the south, exceeding the national average by far. Here agriculture is closely linked to forest activities. Animal husbandry and dairy farming are the most important sectors. Despite the lowland character of the north of the region, soil quality is rather poor due to dry and rocky limestone topsoils. Agricultural structure is small and medium-sized, and management systems quite extensive. The region enjoys high environmental quality in terms of pristine nature, cultural landscapes and biological diversity. The area is also the main water source for Vienna, supplying fresh spring water by pipeline for more than 100 years. Vast areas are subject to restrictions for drinking water conservation and preservation. See also Chapter 7 for further investigations into this case study.

The Neunkirchen basin as the main settlement and economic region is traditionally a core industrial area of Austria dominated by huge manufacturing works and heavy industries. During the last 20–30 years many traditional industries had to close down in the process of restructuring and privatisation. This has caused severe problems of rising unemployment.

Tourism concentrates in a few sites and on a quite extensive level. The main reasons are that young people prefer more crowded places with a wider range of facilities, and on the other hand long-distance travelling has become in vogue due to cheap flights. It seems that improved individual mobility and a better road network have also had a crucial impact on tourism. Tourists stay for a shorter period and overnight stays have become rarer compared with some decades ago. A shift to day-trip and weekend tourism is evident.

After this concise introduction to the socio-economic situation of the case study region, we are going to discuss the particular role and importance of social capital for the socio-cultural micro-level of a community in a remote place endangered by virtual marginalisation, in applying Putnam's social capital concept of association, trust and norms (Putnam, 1993, 2000). For that purpose we chose one of the most extreme examples among all municipalities of the case study district and even in Austria. Schwarzau im Gebirge indicates the worst figures regarding socio-economic indicators such as population decline, overageing, out-migration and brain drain. It has lost more than 60 per cent of its population during the last 150 years. In 2001 only 831 persons lived in Schwarzau, signifying a population density of only 4.4 inhabitants per square kilometre, which is similar to regions north of the Arctic Circle in Finland. In 2003, 25 per cent of the population is aged over 60 years. The main problem is that young people who try to get a better education (high school, college or university) are forced to leave the region. And after finishing they will not be able to find an adequate job. Only the less educated remain, which means that the education level is rather low. Schwarzau is geographically isolated compared to most of the other municipalities of the district. During winter after heavy snowfall the main road connections are sometimes closed because of the avalanche risk, which makes daily life and also commuting quite inconvenient.

The number of agricultural holdings has slumped by 40 per cent between 1970 and 1999, and the number of employees in farm businesses even more. Today all farms are exclusively run by family members, and overwhelmingly on a part-time base. Multifunctionality plays an important and recognised role in farming. It can be illustrated by the close relationship between agriculture and forestry. More than 90 per cent of the municipal territory is covered by forests and so farmers are unsurprisingly connected with wood. About 8000 hectares of forests and woodlands, or 40 per cent of the total area of Schwarzau, are owned and managed by the city of Vienna which is also the most important employer in the region, contributing considerably to the local economy. The main interest of Vienna is to safeguard its spring water supply from this region. The provision of local infrastructure still appears to be more or less satisfactory. There is a kindergarten, primary and secondary school, Red Cross station, taxi, bank and a doctor who also runs a small drug store. Restaurants and pubs traditionally had an important function for community life and for connecting people in times of isolation, restricted accessibility and mobility.

Despite those unfavourable socio-economic conditions, community life is still very vibrant. Schwarzau has weak economic and intellectual capital but rather strong social ties. Civic participation in community-based voluntary associations is impressive. The number of traditional associations is extraordinarily high. Each of the 16 associations has between 20 and 50 members, meaning that a lot of people are organised in several associations at the same time. Due to the wide choice of different associations, few persons are not involved in the socio-cultural life of the municipality. In terms of gender distribution we can observe that there are several associations which are more attractive for women than for men, such as the Red Cross, the choir or a painting circle. There are also two groups of voluntary fire workers which are the only associations that do not accept women as members. Most of the young people are generally well integrated in the traditional associations, but some of them also bring in new ideas which are usually backed by the municipal administration.

In terms of trust, people are generally open-minded and assist their neighbours. Families seem to fragment as elsewhere. Processes that lift out social relations from the local contexts of interaction (Giddens, 1990) also take place in Schwarzau. Forty per cent of the women bring up their children alone. The conflicts between the generations have become more and more evident. Girls and women are getting more independent. In spite of the fact that they are still employed less and that the number of commuters is higher amongst men, most of the women now have their own car. They organise themselves in social groups while family life loses importance.

The importance of civicness (obeying rules and norms) is revealed by a unanimous understanding that most people are rather reluctant to obey rules and regulations. The majority act according to a general consciousness of justice. So they oppose rather than acquiesce to a decision when they do not understand its significance. As a matter of fact it was impossible for the municipality council to enforce a regulation prohibiting the mowing of lawns at the weekend or the kindling of campfires.

Another valuable indicator for social capital is political engagement and participation in elections. In local elections the turnout rate of Schwarzau is above the district and province average, whereas Schwarzau's turnout rate in national elections is less than the average level. This is an indication that the people of Schwarzau are particularly interested in their community affairs. But we also have to consider that participation in municipal elections has been dwindling since 1990, which is an indication of the weakening of social capital.

These results provide strong evidence that social capital has an important function for the municipality of Schwarzau. In this case it seems that social capital really can integrate people, counterbalance economic problems and maintain a comparatively sound environment by keeping up land use and cultivation. The municipal administration is very active in promoting civic life. But this is not an indication for a strong social capital. Putnam (1993) pointed out the threat of personalism and patron–clientelism. Even when the intentions of the decision-making persons are much in favour of their community and they want to integrate all local demands and interests in a quite democratic way, the community will become disorganised as soon as they drop out. Strong local ties often coincide with social exclusion of all those who are not able or willing to accept local norms.

Even in the town of Schwarzau im Gebirge we can expect community collapse when the population declines below a certain threshold, when overageing becomes more and more of a problem, when infrastructure becomes thinned out and when economic and cultural capital decline further.

Berettyóújfalu: A Case Study in Hungary

The Hungarian case study was carried out in the Berettyóújfalu area, a NUTS 4 micro-region located in one of the most underdeveloped regions of Hungary, the Northern Great Plain (NUTS 2). The territory covers some 1400 km² and the population amounts to about 70 000 people in four townships and 27 villages. The average population density of the area comprises 50 persons per square kilometre. Berettyóújfalu suffers from very poor soil quality (especially saline soils of very low fertility) and unfavourable climate conditions (dry hot summers, cold winters).

Marginalisation in the micro-region can be characterised by the following main features (amongst others):

- a steady decrease of the permanent population between 1970 and 1998 of up to 28 per cent in some places;
- a large increase in numbers of economically inactive people since 1990;

- the percentage of people employed in agriculture has declined to about 25 per cent;
- a large part of the grassland now remains uncultivated;
- taxable personal income in the micro-region is only 40 per cent of that of the NUTS 2 macro-region;
- the low education level, with only 7 per cent of the inhabitants of the micro-region having higher education degrees; and
- only 66 per cent of the houses are connected to sewage systems, which is far below the average of the macro-region.

According to Putnam's criteria, the whole country would have to be attributed as 'non-civic' in the 1980s. Hungary was a one-party state with the Communist Party paying serious attention to hampering a strong associational life among the population. All bottom-up civic initiatives (environmental protection, sport, tourism and so on) had been controlled by the state party through its umbrella organisations. Almost all children and young people had been forced to become members of organisations controlled by the Hungarian Socialist Workers' Party (such as the Pioneers' Association, Communist Youth Association). Bottom-up structures through self-organised, independent civic and non-governmental organisations, human rights groups, churches and so on were at a very low level.

This is why the already weak social cohesion and innovative ability of the society withered away almost entirely. Private full-time farms only managed 1 per cent of the total agricultural cultivated area during the 1980s. However, as a matter of fact small agricultural enterprises accounted for one-third of gross agricultural output while they farmed on only about 15 per cent of the total agricultural land. This could be achieved only by agricultural production cooperatives and state farms providing cheap input (seeds, fodder, fertilisers, pesticides, machinery and so on) and services to their members, and employees and also assisting in the marketing of their products. The majority of people working on those large-scale farms were effectively specialised wage-workers with fixed working hours (Szabó, 1988).

Large-scale farms (both state farms and cooperatives) tried to compensate for their losses in the field of agricultural production, especially in animal husbandry, by the development of more profitable non-agricultural activities (industrial, food-industrial, trade, service and so on). During the 1980s more and more units of the large-scale farms did increasingly operate as 'profit centres', trying to be as independent as possible, and their managers followed their own interests. They started to consider the owners of small private plots and auxiliary farms as an obstacle to the development of their own big farm businesses. The close connection and relationship of these small owners to the big farms in a large number of ways – not just organisationally, but also economically and socially – started to be ignored. Thus significant property and income disparities among the agricultural actors already existed before the political changes. In most cases these disparities emerged due to good personal connections and trickery, and not due to better and more efficient work. This further weakened social solidarity, which already had weak roots.

The peaceful political transformation at the beginning of the 1990s brought about a multi-party system and parliamentary democracy in a short time, while the societal and economic transition lasted much longer. The 'Great Disruption' did not comply with what most Hungarians had expected. The severest problem was the loss of employment, which was above all tremendous in agriculture, particularly in agricultural cooperatives. The obligation to employ their members ceased to exist. Therefore the latent unemployment inside of cooperatives turned into open unemployment. This crisis is still acute nowadays, at least in animal husbandry. The number of people employed in agriculture decreased to only one-third of that in 1990. The process of economic change of regime meant significant changes in the employment, income and ownership conditions of all agricultural actors to a much higher degree than incited by the previous political changes.

The position of the agricultural professionals weakened significantly in the period of agricultural crisis and dissolution of large-scale agricultural farms after 1990. Thus their influence towards maintaining social cohesion also declined. The period since 1990 could be regarded as the period of original accumulation of capital and the emerging of a market economy. As a matter of fact this development did not support the building and consolidation of social capital; sometimes it even hampered these processes. Social capital had already been destroyed during the period of communist dictatorship which had lasted for several decades. Unfortunately the establishment of a capitalist society was not in favour of strengthening solidarity, trust and tolerance among citizens. A bipolar farm structure has evolved in Hungary. Now, apart from a relatively small number of large-scale farms (mainly corporations), several thousands of medium- and small-scale farmers are earning their living – at least partly – from agricultural activities. Approximately 200 000 cooperatives, companies and private farmers applied for direct payments from the European Union (EU) in 2004.

While the share of foreign capital is very low in agriculture, the majority of food-processing and trading is now in the hands of foreign investors. The large number of small private farmers is certainly a good basis to establish voluntary networks and cooperatives. However, in reality this development is rather slow. The prevalence of the market economy, business connections and the everyday struggle for existence only permits a very narrow framework for the development of a 'strong associational life'. Among people smitten with permanent unemployment, psychological problems are frequently accompanied by different types of organic diseases.

Due to the weakness of 'civic engagement' there is much room for corruption as well as for the black and grey economies in the whole national economy, and also in agriculture. According to the behavioural patterns developed during the past decade, the concealment of taxable income is not at all a socially condemned action in Hungary. It seems to be a paradox, but in the case of a group of small agricultural producers, tax legislation offers the prospect of avoiding paying taxes. It is worth mentioning that, according to a survey of the Hungarian Central Statistical Office, only 20 per cent of the total annual working time is paid in agriculture (HCSO, 2004).

Hungarian agriculture was neglected by the policy-makers and the majority of farmers were not prepared for the Common Agricultural Policy (CAP). This caused a very heavy shock for farmers with small or medium-sized landholdings. The lack of appropriate national agricultural and regional strategies as well as the weakness of local civic cooperation seem to be the prominent obstacles to coping with marginalisation. However, despite the fragmented farm structure, cooperation among farmers (cooperatives, producers' associations and so on) is developing – but rather slowly (Szabó and Kiss, 2004). Due to this situation the Hungarian farmers are at the mercy of multinational food and trade companies (Dorgai, 2005).

We presume that, after a period of transitional distortions, growing potential provided by the CAP (market security offered by intervention measures, increasing rural development subsidies, necessity of developing producers' associations in the vegetable and fruit sectors and so on), will also bring about positive changes concerning social capital (Szabó and Fertő, 2004). In this respect the most rapid and comprehensive outcomes can be achieved by the LEADER programme.

Weakness of social capital can be traced in the weakness of human resources and the low degree of willingness to cooperate. It is interesting to mention that during the referendum concerning the accession to the EU in 2003 the turnout rate was far below the national average in the case study region. Some characteristics of weak social capital in the micro-region are as follows:

- The weakness of human capital is revealed by the low proportion of people with higher educational degrees, and in overageing.
- The willingness to cooperate is weak, despite the fact that half of the population aged between 15 and 60 are engaged in small and medium-sized agricultural farm activities. When based on the calculation of annual work units (AWU), these farms tie down 20 per cent of the full working time of the population.
- Fruit and vegetable production plays a significant role in the micro-region. As part of the CAP, fruit and vegetable producers are supported solely through producers' associations. However, the establishment of such organisations is slowly going on in the micro-region.

In February and March 2004 we held lectures and consultation sessions entitled 'Expected Impacts of the European Union's Agricultural Policies on

Hungarian Agriculture' for the farmers of the 13 settlements in the Hajdú-Bihar county (NUTS 3). On average 15–70 persons took part in these events. Another workshop was organised for the stakeholders of the micro-region (farmers, mayors, consultants, managers of food companies, staff members of the University of Debrecen and so on), in September 2004. Regarding their questions and remarks it became obvious that the willingness for farm cooperation is low, which is due to a general mistrust. It is worth mentioning that the participants complained about insufficient information and the lack of professional advice. They feel defenceless against the processing and purchasing companies which are in a monopolistic position.

LEADER programmes could play a significant role in strengthening social capital in Hungary and perhaps in all ten member states that entered the EU in 2004. They are indispensable from the perspective of bottom-up initiatives and successful development of the micro-region. Therefore it is important to mention that in the new member states at least 2.5 per cent of budget estimates of the European Agricultural Fund for Rural Development are obligatorily assigned for that purpose in the period 2007–13 (*Official Journal*, 2005).

Mäntyharju: Agricultural Marginalisation Combated by Local Activism in Finland

Chapter 5 of this volume showed that agriculture was diminishing in large areas of eastern and northern Finland. Usually agricultural land is turned into forest, either in a controlled process of reforestation or by natural processes.

The present economic circumstances and policy make farming economically unviable. However, agriculture contributes to rural life and is an important form of land use which preserves cultural inheritance. Since fields comprise only 8 per cent of the land area, their role in creating open landscape is crucial and small changes in the field area can have large local impacts.

Vulnerability of agriculture is appropriate in the Finnish case since agriculture is almost totally policy dependent. The harsh climate and short growing period result in a low general level of production per unit, although the level of technology development is very high. The whole area of Finland is classified as a Less Favoured Area (LFA). The most profitable line is dairy production. The productivity of animals is comparable to that of Central Europe, but since the expenses are much higher because of the long winter, the economic result is heavily dependent on the EU and nationally financed support systems. The economic base of Finnish agriculture is slowly eroding.

Agriculture in Finland has undergone a very rapid structural change. This process started in the 1970s but has accelerated with membership to the EU. The number of farms has decreased dramatically and the average cultivated area has increased. Many Finnish farms are pluriactive with business diversification on- or off-farm, or wage work outside the farm.

The municipality of Mäntyharju in south-eastern Finland has many characteristics of an area prone to marginalisation. In the Finnish rural typology it belongs to the sparsely populated countryside, having seven inhabitants per square kilometre. The land use statistics show a dramatic decrease in the field area during the 1990s and the municipality comes under the Objective One programme for Eastern Finland.

The marginalisation processes defined by the biophysical and socio-economic indicators are evident in Mäntyharju. There is a clear trend towards a declining number of farms and a decline in utilised agricultural area. The population is ageing and the income level is far below the national average. The population is concentrated in Mäntyharju. With the closing down of farms, smaller villages turn into dormitories. The villages become quiet places with few hands to keep up the houses and the surroundings.

The prevailing rurality and the importance of agriculture are obvious. The share of people employed in agriculture and forestry is 14.4 per cent of all employed persons, compared to the national average of 4 per cent. The cultivated area has decreased by 47 per cent between 1969 and 2000. The abandoned fields have been reforested either deliberately or by natural process. In 2003 the average farm size in Mäntyharju was 18.6 hectares of fields, while in the whole country it was 31 hectares. Farmers in Mäntyharju earn on the average 17 per cent less than their colleagues in the county of South-Savo, and 26 per cent less than their colleagues in the whole of Finland. The reasons are a change from dairy to grain production due to changes in support levels for dairy with the accession to the EU, and the geographical obstacles for increasing the cultivated areas of the existing farms.

During recent years hardly any elderly farmer has had a successor. This fact reveals uncertainty about the future in farming due to changing policies and short perspectives. Agriculture in Mäntyharju will continue to diminish if the generation change frequency does not increase considerably.

This clear trend towards agricultural marginalisation is, however, not observed in the commonly used socio-economic indicators. Most importantly the long-term trend of population decrease in Mäntyharju has turned. Gender distribution is balanced, with no evidence of stronger female out-migration. The unemployment rate is less than 10 per cent, which is close to the national average and far less than in South-Savo and eastern Finland.

The overall structure of Mäntyharju industries is relatively favourable. Mäntyharju has a very high level of employment self-sufficiency due to its considerable industrial base – 35 per cent of all jobs are in the industrial sector. In 2002, there were 111 small enterprises employing 171 persons in Mäntyharju (Statistics Finland, 2004). The self-sufficiency in jobs is 95 per cent, but since there is commuting both from Mäntyharju to other areas and from other areas to Mäntyharju, some 14.5 per cent of the workforce commutes out of Mäntyharju.

A specific feature in Mäntyharju is undoubtedly the great number of second homes. The number has been steadily increasing and is around 4500 in 2005. The summer guests double the population of Mäntyharju for some 86 days a year. The economic impact of summer guests is clearly visible in the number of shops in Mäntyharju municipality centre. But otherwise, financial flows are not that great because many leisure-time activities are free; in the Nordic countries everyone is allowed to wander in forests and pick berries or mushrooms, for example, and for hobby fishing there is only a nominal fee. Rural amenities are difficult to sell in Finland, where there is a lot of space and a long tradition of common use of natural resources.

The marginalisation analysis of Mäntyharju gives a mixed message. Traditional activities like farming and forestry are declining; the population is ageing and concentrating in the centre. However, there are many village associations and Mäntyharju is a part of an LAG that finances development efforts by various associations, groups, firms and private people. These local-level development efforts indicate the presence of considerable social capital and also the process of an increasing level of social capital.

The 11 village associations have social activities and usually common local premises for meetings, hobbies and celebrations. There are activities targeted at the protection of the local environment and small development projects. Some village associations also have Internet sites providing information about local entrepreneurs, hobby and sports activities and cultural activities. All the village associations belong to the regional village network association, called *Järvi-Suomen kylät ry*. This network organises training and marketing activities where all the village associations of the county take part. It acts also as a pressure group towards the regional and national administrations.

In Finland LAGs are mainstreamed in the national rural policy. They cover the whole country except for the central areas of some major cities. In Mäntyharju there are currently nine development projects by the Veej'jakaja, the LAG under the national POMO+ programme which is similar to LEADER+ but completely nationally funded. The action group projects have supported rural industries development in Mäntyharju. They have also increased cooperation between villages and strengthened the old *talkoo* heritage, which means gatherings for voluntary work around a well-defined goal. The long-term village movement and new LAGs with their economic resources have built social capital which is recognisable though difficult to measure.

In conclusion there are powerful indicators of marginalisation as a part of the overall marginalisation of Eastern Finland and especially of sparsely populated rural areas. Local dynamism is a strong counterforce, partly because of the lively tradition of association life in Finland coupled with the village revival. These are spontaneous endogenous developments which strengthen and create social capital. The regional level association of villages and the LAG have also accumulated political capital which is used to secure resources at local, regional and national level for bottom-up development. This trend is greatly strengthened by the innovative EU policy of LEADER+ programmes, which has been mainstreamed in Finland probably because of the strong rural association tradition and the growing social and political capital due to LEADER+ programmes.

In Finland, the importance of social capital was recognised in 1998 in the report of the Special Parliamentary Committee for the Future. In this document a special section is devoted to social capital focusing on the importance of education, as well as productivity and equality, from the standpoint of the potential of individuals. Alanen and Pelkonen (2000) concluded in their study on Finnish social capital that there is a detectable and consistent statistical association between social factors and regional growth. Participation in leisure organisations (like village associations) showed statistical correlation with economic growth.

DISENTANGLING SOCIAL CAPITAL: CONCLUSION AND DEBATE

What particular conclusions can be drawn from the experiences of the three case studies? Can the concept of social capital explain the sometimes loose correlation between efforts and successful results – that is, the number of policy measures applied and the state of development in a region? Can social capital compensate for structural deficiencies? And what are the roles of multifunctionality, local governance and the perception of territorial civicness in this game of gaining and maintaining economic, socio-cultural and environmental sustainability?

We have applied the concept of multifunctionality not only with regard to agricultural multifunctionality but also in the broader sense of any economic activity in rural areas generating positive and negative externalities, connecting people and resources. Multifunctionality made visible may broaden the realm of opportunities in utilising local assets and in preventing negative impacts on local society and nature. Social capital thrives more easily under sound economic socio-cultural and environmental conditions. As Putnam (2000) announced, a 'well-connected individual in a poorly connected society is not as productive as a well-connected individual in a well-connected society. And even a poorly connected individual may derive some of the spill-over benefits from living in a well-connected community.' This indicates thoroughly the limits and restrictions of social capital as a tool for rural development. Where the preconditions are poor, damage is likely to occur and social capital could hardly compensate, or at least extraordinary efforts are needed. Simultaneously, social capital is a precious asset. A connected society that is rich in social capital may promote rural development more easily.

Anyhow, as seen in the examples presented in this chapter, the quantity of existing local social capital may play a decisive role but its influence should also not be overestimated. Social capital is not likely to solve all the problems alone. It facilitates the utilisation of local resources both in terms of natural and human resources by the creation of social networks, trust and civicness. Besides those positive aspects there are also some negative tendencies that have to be considered.

Global economic transformation has an important impact as well. In the era of globalisation, rural areas and their populations are subject to vast socio-economic transformation with a strong impact on the whole fabric of local communities. Many processes and forces prove detrimental to civic engagement. People have less time and leisure for voluntary associations; TV, telecommunication and Internet produce a virtual neighbourhood. People nowadays do not necessarily need to link to each other to gain sufficient goods, entertainment and information. Through loss of local infrastructure, and the closing down of local shops and restaurants, gathering places are also becoming sparse and thus opportunities for personal contact have decreased.

Commuting brings about a spatial fragmentation between home and workplace which in the long run might be bad for local voluntary associations and community life. But at the same time commuting also fosters the bridging of social capital. Putnam (2000) distinguishes between bridging and bonding social capital. Bonding social capital is good for undergirding specific reciprocity and mobilising solidarity. Dense networks in ethnic enclaves for example provide crucial social and psychological support for less-fortunate members of the community. Bridging networks, by contrast, are better for linkage to external assets and for informal diffusion. Weak ties that link a person to distant acquaintances who move in different circles from their own are actually more valuable than strong ties that link them to relatives and intimate friends whose sociological niche is very likely the person's own. Bonding social capital, by creating strong in-group loyalty, may also create strong out-group antagonism. People who work outside their region are exposed to a wider array of social and community networks facilitating new and ample experiences. Hence, through them innovations and new ideas from outside may also come more easily into rural regions.

Networks and associated norms of reciprocity are generally good for those inside the network, but the external effects of social capital are by no means always positive. Social capital can probably also have a dark side when people become excluded who do not want to join or accept the dominant local norms. Community connections are sometimes oppressive. Such persons tend to become outcasts and are socially excluded or marginalised, maybe more easily than in anonymous town life. The greater homogeneity of a community, the more difficult it becomes for incomers to be accepted. Communities with high rates of residential turnover are maybe less well integrated but also more tolerant.

There are some points that should be acknowledged when critically debating and interrogating social capital:

- Putnam's concept of social capital has to be modified in some regards. The number of associations per inhabitants which has been used for measuring social capital (Putnam, 1993) tells very little about the kind of the associations and whether they are appropriate to address and encompass all local people. Sometimes only a few persons are involved in a great number of associations while many other groups are not concerned (for example women, youth, ethnic and religious minorities).
- This fact calls for gendering social capital. The distribution of social capital between men and women might be unequal, like the distribution of material (financial) and human (intellectual, cultural) capital. In many European rural regions women became the majority of permanent dwellers at least during the week when their husbands are commuting. They are developing their own female-dominated networks. In other regions such as in Northern Europe, the opposite is the case: women work outside the farm while men continue farming. Then women migrate more easily and leave men behind.
- Social capital may not be considered as a constant and stable feature. It is generally in flux and transition. The structure of social capital has to adapt to new challenges and developments. New collective organisations will have to emerge in response to new needs.
- Some government policies almost certainly have the effect of destroying social capital. For example, the closing down of railroads, post offices and public services are disrupting existing community ties. The local civic community can hardly replace or compensate for the deficiencies generated by state withdrawal. Rural development needs hardware for institutional infrastructure. Social capital cannot exist in a void, and as Putnam (2000) mentioned, 'social capital is self-reinforcing and benefits most those who already have a stock in which to trade'.
- Social capital can also be promoted by government action creating institutional structures that encourage cooperation and give opportunities for learning and thus increased trust between the local actors. LEADER programmes are a case in point. In other terms this kind of action increases the structural political capital since it influences the possibilities of diverse actors to accumulate political capital.
- Particularly in the EU member states of Eastern and Central Europe, the discussion about social capital is of crucial importance. During the former socialist regimes traditional ties and connectedness between rural people weakened through agricultural collectivisation and industrialisation. The

political transition at the beginning of the 1990s destroyed the collective socialist structures, but at the same time the traditional social structures could not be regained and thus social capital is quite feeble. In this case supporting social capital by fostering associations and mutual trust could be an important means of countering rural marginalisation.

Since our analysis points out clearly the importance of social capital in rural development dynamics, we suggest that social capital should be more recognised by policy-makers as a key factor in the rural development process, hampering (when weak) or helping (when strong and well rooted) the implementation of rural development policies and specifically those policy measures aimed at counteracting marginalisation processes and, to some extent, land abandonment. Thus creation of social capital should be a conscious aim of any rural development policy. But also, the dark side of social capital such as excluding incomers and obstructing innovation must not be forgotten. Moreover, one should be aware that social capital is just a software package for rural development besides others. We also need to have a functioning hardware in terms of local employment, infrastructure and services. Cutting down those facilities would mean this software cannot be installed successfully any more.

REFERENCES

Alanen, Aku and Lea Pelkonen (2000), 'Can regional economic growth be explained by social capital?' in Jouko Kajanoja and Jussi Simpura (eds), *Social Capital, Global and Local Perspectives*, Helsinki: Government Institute for Economic Research, pp. 51–76.

Árnason, Arnar, Jo Lee and Mark Shucksmith (2004), *Understanding networks and social capital in European rural development, Improving Living Conditions and Quality of Life in Rural Europe*, EU Conference Westport, Ireland, May 2004.

Baumann, Pari (2000), 'Sustainable livelihoods and political capital: arguments and evidence from decentralisation and natural resource management in India', London: Overseas Development Institute, Working Paper 136.

Birner, Regina and Heidi Wittmer (2003), 'Using social capital to create political capital: how do local communities gain political influence? A theoretical approach and empirical evidence from Thailand', in N. Dolsak and E. Ostrom (eds), *The Commons in the New Millennium: Challenges and Adaptations*, Cambridge, MA: MIT, pp. 3–34.

Booth, John A. and Patricia Bayer Richard (1997), 'Civil society, political capital, and democratization in Central America', Paper presented at the

XXI International Congress of the Latin American Studies Association 17–19 April, Guadalaraja, Mexico.

Bourdieu, Pierre (1979), *La distiction. Critique sociale du jugement*, Paris: Édition de Minuit.

Bourdieu, Pierre (1986), 'The forms of capital', in John Richardson (ed.), *Handbook of Theory and Research for the Sociology of Education*, New York: Greenwood Press, pp. 241–58.

Burt, Ronald S. (2000), 'The network structure of social capital', in Robert I. Sutton and Barry M. Staw (eds), *Research in Organizational Behavior,* Greenwich, CT: JAI Press, pp. 345–423.

Coleman, James S. (1988), 'Social capital in creation of human capital', *American Journal of Sociology*, 94, 95–120.

Dorgai, L. (ed.) (2005), 'Termelői szerveződések, termelői csoportok a mezőgazdaságban' (Producer organisations, producer groups in agriculture), *Agrárgazdasági Tanulmányok*, Vol. 4.

Fukuyama, Francis (1999), 'Social capital and civil society', Paper prepared for the International Monetary Fund conference on Second Generation Reforms, November 8–9, Washington, DC.

Giddens, Anthony (1990), *The Consequences of Modernity*, Oxford: Basil Blackwell.

Hungarian Central Statistical Office (HCSO) (2004), *Statistical Yearbook of Agriculture 2003*, Budapest: Hungarian Central Statistical Office.

Norris, Pippa and Ronald Inglehart (2003), *Gendering Social Capital: Bowling in Women's Leagues?*, Cambridge, MA: Harvard University Press.

Official Journal of the European Union (2005), 'Council Regulation (EC) No 1698/2005 of 20 September 2005 on support for rural development by the European Agricultural Fund for Rural Development (EAFRD) L 277/1-40', Volume 48, 21 October.

Putnam, Robert D. (1993), *Making Democracy Work: Civic Traditions in Modern Italy*, Princeton, NJ: Princeton University Press.

Putnam, Robert D. (2000), *Bowling Alone: The Collapse and Revival of American Community*, New York: Simon & Schuster.

Rakodi, Carole (1999), 'A capital assets framework for analysing livelihood strategies', *Development Policy Review*, 17 (3), 315–42.

Schubert, Robert (2005), 'Farming's new feudalism: consolidation and biotechnology shrink farmers' options', *World Watch Magazine*, 18 (3), 10–15.

Szabó, Gabor (1988), 'Élelmiszer-gazdaságtan' (Food economics), Lecture notes, Kaposvár: Pannon Agricultural University.

Szabó, G.G. and I. Fertő (2004), 'Issues of vertical co-ordination by co-operatives: a Hungarian case study in the fruit and vegetable sector', in J. Berács, J. Lehota, I. Piskóti and G. Rekettye (eds), *Marketing Theory and*

Practice: A Hungarian Perspective, Budapest: Akadémiai Kiadó, pp. 362–79.

Szabó, G.G. and A. Kiss (2004), 'Economic substance and legal regulation of agricultural co-operatives in Hungary', in Carlo Borzaga and Roger Spear (eds), *Trends and Challenges for Co-operatives and Social Enterprises in Developed and Transition Countries*, Trento, Italy: Fondazione Caríplo, Edizioni 31, pp. 265–80.

Tillberg-Mattsson, Karin and Susanne Stenbacka (2004), *The Role of Social Capital in Local Development. The Case of Leksand and Rättvik, Sweden*, National Report to Restructuring in Marginal Rural Areas (RESTRIM), Uppsala: The Institute for Housing and Urban Research.

Woolcock, Michael (1998), 'Social capital and economic development: toward a theoretical synthesis and policy framework', *Theory and Society*, 27, 151–208.

12. Understanding the quality of land in agricultural land use systems

Anna Martha Elgersma, Shivcharn S. Dhillion, Arnold Arnoldussen, Josef Fanta and Eva Boucníková

In the recent decades agricultural land cover and its designation have undergone significant change in Europe. The understanding and using of land as an arena that should have provisions not only for production has been gaining ground gradually, and has transpired into the reconfiguration of land utility and function. How land is to be managed and the productivity it ought to display in economic terms as criteria for its continued and defined use is based on efficiency, specialisation and maximising production. A shift from agricultural systems adapted to soil characteristics, and physical and biological components of land, to agricultural systems not adapted to soil and land has taken place. This shift has occurred mainly through intensification and extensification needs.

The processes of intensification and extensification have been triggered through policy-driven initiatives, which have varied during the twentieth century. In some cases the results have been devastating to society, leading to almost permanent loss of productivity, high soil rehabilitation costs, increasing proneness to erosion, loss of soil organic matter, and groundwater and surface water pollution (Burt, 2001; Morari et al., 2006). In fact, in Southern Europe soil degradation of agricultural land through inadequate management has become a very serious problem. Criteria for the designation of land use have thus not always had a sound rationale, taking into account both short- and long-term consequences. The issue of the quality of land is vital, in that it can define how land is to be used. The quality of land, or land quality, refers to the condition of land relative to the requirements of land use, including agricultural production, forestry, conservation and environmental management (Dumanski and Pieri, 2000: 93). According to Wiebe (2003: 2):

> Land quality refers to the ability of land to produce goods and services that are valued by humans. This ability derives from inherent/natural attributes of soils

(e.g. depth and fertility), water, climate, topography, vegetation, and hydrology as well as 'produced' attributes, such as infrastructure (e.g. irrigation) and proximity to population centers.

In essence quality of land has local attributes, which are both qualitative and quantitative.

For the purposes of this chapter we will focus on the soil, terrain and biological aspects of land quality. In this chapter we argue that the physical and biological aspects of land and its soil have been given little priority in the designation of land use, and that inadequate understanding of these land quality aspects can lead to serious burdens on society. To illustrate this we present case studies through history up to the current time. We explore traits and indicators of different criteria of land types and address two salient developments (Baldock et al., 2002) in agriculture:

- intensification on suitable land in favourable economic situations; and
- marginalisation on less-suitable or unsuitable land in economically marginal situations.

Prior to addressing the above we present a brief review of soil quality to anchor the issue of suitable versus unsuitable land.

SUITABLE VERSUS UNSUITABLE LAND

Suitable land for agriculture has soils, terrain and climate conditions favourable for agriculture. The land is easily accessible and cultivatable. Suitable land offers better economic outcomes for agriculture than less-suitable or unsuitable land. Less-suitable or unsuitable land has soils which are less or not favourable for agriculture. Also the climate and terrain conditions can be unfavourable and the accessibility and agricultural potential of the land is restricted.

The suitability of soils for agriculture can be measured by soil properties and their indicators. Those can be divided into static and dynamic properties and indicators (Carter, 2001). The static ones are those such as texture (particle size distribution), soil depth and rooting depth, stoniness, water-holding capacity and the composition of minerals. The dynamic ones are, for example, soil structure, soil organic matter content and soil biodiversity. The dynamic properties can change rapidly when land use changes take place (Carter, 2001; Stockfisch et al., 1999). The changes in land use – particularly by intensification, scale enlargement and marginalisation of agriculture – in the past half-century have had a tremendous impact on the soil, especially its dynamic properties. In many cases it has caused soil degradation.

According to the European Commission (2002) soil degradation is indicated by soil erosion, decline in organic matter, soil sealing (change in the nature of the soil such that it behaves as an impermeable medium, for example due to compaction by heavy machinery), soil contamination, decline in soil biodiversity, salination and sodification, and landslides and floods. According to the European Commission soil degradation threatens the functions soil has for society. It deteriorates the productivity of soils, the soil as a habitat and gene pool of soil organisms, the regulation of water and nutrients, the soil as a platform for human activity, and the soil as a functional element of landscape and cultural heritage. Both soil functions and degradation are related to land quality. For agriculture, soil productivity is crucial for the viability of agriculture. Soil productivity is closely related to the habitat and gene pool of organisms, the regulation of water and nutrients, and the terrain conditions. Soil degradation decreases soil productivity, water and air availability, and rooting space. In the worst case, soil degradation can lead to marginalisation of agricultural activity as degradation leads to lower production levels, perhaps rendering it not viable economically.

LAND USE CHANGES IN THE AGRICULTURAL LANDSCAPE

Historical Land Use

Ever since the Neolithic Period (the New Stone Age, traditionally the last part of the Stone Age which began around 8500–8000 BC, and part of the period in the development of human technology) and the rise of settled agriculture, there was an ever-increasing need for more agricultural land to feed the growing population. Land was reclaimed from natural areas such as forests, wetlands or other natural ecosystems. Lowlands, coastal areas, broad river valleys and transition areas between forest and steppe were used first for agriculture (Küster, 1999). Natural habitats were replaced by man-made ecosystems used for agriculture, forestry or grazing. During the first period, shifting cultivation was commonly practiced. But not long after, permanent land use became prominent. Methods of land use and soil cultivation were invented and applied to keep land fertile and in production. Soil exhaustion was counteracted by organic matter supplied in the form of dung and/or forest litter to compensate for the decrease of nutrients, or to improve soil fertility like the development of anthrosols on poor sandy soils in Flanders, the Netherlands and Lower Saxony (Spek, 2004). Soil fertility was maintained by, among other techniques, crop rotation systems, drainage and irrigation, in combination with periodical organic dung input and the practice of fallow. Where land was scarce and/or climate conditions extreme, various forms of transhumance developed between lowlands and mountain areas. Farming practices were driven by soil and climatic conditions and scarcity of land. Agricultural land use was fully adapted to the levels of soil fertility, slope conditions and microclimate.

In periods of concentrated and targeted colonisation in the Middle Ages (the period between the fifth and fifteenth centuries), taking European uplands into culture became a complicated process. Due to excessive erosion of land after the clearing of upland forests, the fourteenth century is known as the period of largest man-induced relief changes in the European Holocene (Bork et al., 1998). Overgrazing of heathlands in combination with climatic fluctuations changed North-West European heathlands into extensive blown-sand areas. Land use in erosion-sensitive areas was subject to specific rules. Construction of terraces on slopes and/or stone walls and hedgerows served the main task of protecting the land against erosion, and simultaneously created a well-considered land and landscape organisation. Supported by limited but effective technical measures, these systems represented the usual practice of the pre-industrial period, applied for centuries. The terrace culture in Southern Europe is a good example of this. Such traditional farming systems created a heterogeneous agricultural landscape characterised by locally typical patterns of fields and grasslands, divided by baulks, hedgerows or ditches, with specific uncultivated wet or dry habitats and a spontaneous shrub and tree vegetation. These linear and/or small-scale landscape elements represented valuable habitats for plant and animal species and were the source of a high landscape and biological diversity.

Cultivable landscapes with extreme natural conditions – for example steppes, mountain areas, alluvial wetlands or extensive sand covered areas with poor or shallow soils – often allowed only for extensive land use. Due to this, natural and environmental conditions of these areas were only slightly changed and specific landscape features were maintained. When the unsuitability of land for agriculture was not taken into account for management, this led to the deterioration of soil, land and landscape. For instance forest clearing and overgrazing over centuries has caused severe erosion in Iceland (Arnalds, 2005) and in the Mediterranean region.

Current Land Use

The current use of agricultural land
At present a large proportion of the land in use is not adapted to the quality of the land, as land use is often determined by the type of management and agriculture practice and not suitability. This is due to the management and practised agricultural systems. According Reidsma et al. (2006) the current agricultural systems can be classified in the following categories:

- extensive grassland management;
- extensive organic farming;

- extensive farming;
- intensive organic farming;
- intensive grassland management;
- highly intensive organic farming;
- intensive production systems; and
- highly intensive production systems.

In Reidsma's classification each system received a value for ecosystem quality, measured by the relationship between the farming type and biodiversity. Extensive grassland management had the highest value for ecosystem quality while the highly intensive production systems got the lowest value. The low values for the intensive systems are due to the fact that such systems do not consider land quality as an essential part of the management, particularly not in relation to the degradation of soil fertility, soil structure, soil biodiversity, soil organic matter content, water availability and valuable habitats established by previous land use systems, as well as the landscape mosaic.

The degradation of the soil is linked largely to dynamic soil properties (e.g. porosity and retention). Such properties represent the degradation of the soil functions as defined by the European Commission (2002). The increase of soil degradation by erosion and desertification but also by soil compaction, acidification, salination and a decline in soil organic matter leads to a yield loss (for example Håkansson and Medvedev, 1995; Hajkowicz and Young, 2005; Arriaga and Lowery, 2003; De la Rosa et al., 2000; Bakker et al., 2004; Ball et al., 1997). In the worst case it leads to marginalisation of agriculture.

A main reason for soil degradation is that a large part of land reclaimed for agriculture is not suitable for intensification practices, or for more efficient management through area enlargement and using heavy machinery, or for changes from permanent crops into arable fields and specialisation. This is illustrated by Table 12.1, showing that 52 per cent of the agricultural land of Southern European countries has low quality for agricultural use due to restrictions in soil, terrain and climate for agriculture. According to Table 12.1 only 14 per cent of the agricultural land has high quality for agricultural use, which varies from 8 per cent for Portugal to 31 per cent for Italy, and could be classified as favoured areas for agriculture. Probably, most of the areas with moderate or low quality for agriculture and areas without data (Table 12.1) are in areas marginal for agriculture and are classified as Less Favoured Areas (LFAs). For instance 86 per cent of the utilised agricultural area (UAA) in Portugal and 80 per cent in Spain is classified as LFAs and falls into areas with special 'handicaps' for agriculture. These handicaps are, among others, steep slopes, remote fields, and stony, shallow, infertile and dry soils.

Table 12.1　Quality of land (indicated by soil, relief, climate) for agriculture

Country	High quality km² (%)	Moderate quality km² (%)	Low quality km² (%)	No data available km² (%)
France (southern part)	10 456 (6)	51 546 (27)	85 690 (45)	42 469 (22)
Italy	93 351 (31)	78 672 (26)	117 772 (39)	11 483 (4)
Greece	24 919 (19)	23 394 (18)	75 775 (57)	7 903 (6)
Spain	35 286 (7)	149 026 (30)	292 586 (59)	20 619 (4)
Portugal	7 214 (8)	22 236 (25)	58 475 (66)	1 003 (1)
Total	171 226 (14)	324 874 (27)	630 298 (52)	83 477 (7)

Source: Data for South European countries Programme CORINE 'Soil Erosion Risk and Important Land Resources'.

Degradation of land quality is also apparent in a degradation of valuable habitats. Agriculture is responsible for a considerable part of the decline of habitats linked to traditional agricultural practices. This is illustrated in Table 12.2 where the decline of dry or mesic grassland habitats is shown. Agriculture and forestry represent a decline of between 28.9 and 71.9 per cent for the countries in Table 12.2. But soil sealing by urbanisation, industry, mining, tourism and so on also has a considerable impact on the decline of these habitats. In general, loss of grassland habitats is a major concern for biodiversity conservation (Young et al., 2005).

The more extensive organic and traditional farming systems are more adapted to the quality of land. These farming systems are used by farmers who cautiously use mineral fertilizer and pesticides, employ so-called 'good farming practices', and contribute to enhancing and maintaining soil productivity and the conservation of organic matter, soil structure, water and nutrient regulation and the soil as a habitat and gene pool of soil organisms. Extensive farming systems only occasionally exceed the available soil productivity and conservation aspects of soil. In organic farming, managing soil fertility without fertilisers and pesticides is a main aim (Watson et al., 2002; Lægreid et al., 1999). Organic farmers rely on the management of organic matter which improves, particularly, the dynamic soil properties. However, depending on the soil properties and climate, organic farmers need to be cautious not to destroy soil structure by tillage and using heavy vehicles under wet conditions, thus avoiding soil compaction (Pulleman et al., 2003).

With respect to the link between land quality and agriculture, a special case is traditional farming. Traditional farming is based on local traditions

Table 12.2 Main impact of land use and activity types (%) in lowland areas with more than 30% dry or mesic grassland habitat coverage in potential sites of the European Community interest under the Flora, Fauna and Habitats Directive (pSCIs)

Country/ number of sites	Agriculture, forestry	Hunting and collecting	Human induced changes in hydraulic conditions	Leisure and tourism	Mining and extraction of materials	Natural processes	Pollution and other human impacts	Transportation and communication	Urbanisation, industrialisation and similar activities
Austria 6	71.9	6.3	3.1	6.3	3.1	3.1	0.0	3.1	3.1
Finland 8	47.1	0.0	0.0	5.9	5.9	17.6	5.9	0.0	17.6
France 131	40.1	15.8	3.4	13.6	2.3	5.7	4.4	7.7	6.9
Greece 11	29.7	21.7	8.6	5.7	2.9	8.0	4.0	12.6	6.9
Ireland 6	39.4	6.1	0.0	3.0	12.1	9.1	3.0	21.2	6.1
Italy 391	46.6	11.9	3.1	6.8	2.2	3.4	2.9	15.8	7.4
Portugal 11	38.0	10.9	5.8	10.2	3.6	8.8	10.9	5.8	5.8
Spain 144	28.9	12.8	3.7	16.2	5.6	7.9	6.4	11.0	7.5
Sweden 147	64.4	2.2	0.0	6.7	0.0	8.9	15.6	2.2	0.0

Source: Data as reported in Natura 2000 forms by the end of 1999 (EEA, 2001: 10).

and often contributes to maintaining valuable agri-ecosystems. Many of the traditional farming systems can be considered as 'high nature value' (HNV) farming systems. HNV farming systems play an important role in preserving biodiversity, habitats, soil productivity and the above-mentioned conservation aspects. In most member states of the European Union (EU), these farming systems cover 10 to 25 per cent of the agricultural area. HNV farming systems are one of the EU-level priorities 'to protect and enhance the EU's natural resources and landscapes in rural areas' (Commission staff working document 2005: 3, 9). A major proportion of HNV farming systems form a part of the LFAs in uplands. During the 1990s, the largest changes in land use are in LFAs with valuable agri-ecosystems. One reason for this is that productivity of such systems may be low and they are labour intensive and may be prone to marginalisation. For instance, in Spain the terraced hillside cultures on less-suitable or unsuitable soils have been abandoned in favour of cash crop cultivation on more suitable soils on the plains, obtaining much higher net outputs (Puigdefabregas, 1998).

Size of utilised agricultural area (UAA)

Land quality can be a deciding factor in land to be taken out of production. Due to the intensification and efficiency in agriculture in the second half of the twentieth century, the UAA declined. Less-suitable and unsuitable land was taken out of production following the set-aside policy of the Common Agricultural Policy (CAP) of the European Union. Crop yield, economic considerations, soil type and drainage mostly influenced the choice of land to be set aside (Firbank et al., 2003). The decline in UAA was also caused by afforestation of land, stimulated by the Afforestation Regulation 2080/1992 of the CAP and by the rural development policy of the CAP (Council regulation (EC) No 1257/1999). It mostly took place on less-suitable and unsuitable land for agriculture. Furthermore the decline in UAA is caused by abandonment of land due to marginalisation of agriculture and is mainly related to less-suitable or unsuitable land. Land with low soil fertility or a sensitivity to drought, land with stony or shallow soils, badly drained soils, degraded land, and land with less-suitable or unsuitable terrain with small fields, or less-accessible and remote land, is taken out production first (Flinn et al., 2005; Puigdefabregas, 1998; Nertsen et al., 1999; Kosmas et al., 2000; Duram et al., 2004; Mottet et al., 2006; Chapter 5 of this volume). In other cases land quality as related to agriculture is not a decision factor for land taken out of production. Soil sealing, that is, land used for the expansion of settlements, industrial areas, sports fields and so on, does not consider land quality as a decision factor. Soil sealing may even cause the closing down of farms with suitable land. Soil sealing is at present one of the main threats to soil functioning, which include nutrient and water retention capacity, porosity, and fertility (Gunreben, 2005). Agricultural land taken out of production destined for nature development as part of Ecological Networks (Natura

2000) also occurs on suitable and unsuitable land. Another example is that suitable and unsuitable land was abandoned in Eastern Europe due to marginalisation of agriculture after the collapse of the Soviet period (Chapter 6 of this volume; Meier et al., 2005; Peterson and Aunap, 1998).

Abandoned land

Abandoned land requires special attention when it causes degradation of the soil due to unfavourable land cover changes and a lack of management. The point is that degraded land has low value and has restricted use as other land use types like forests or agriculture.

The impact of abandonment on soil degradation varies across Europe. The development of soil depends on both geographic and climatic conditions, vegetation development and disturbance of vegetation development (Table 12.3). In the case of semi-arid regions abandonment of agricultural land is usually followed by degradation of soil functioning (Bautista et al., 2004) due to erosion and desertification. The main reason for this is that the soil of arable land is left uncovered after abandonment and erosion can take place. The risk of fire increases where vegetation develops. The risk of fire increases rapidly as fuel accumulation and landscape homogenisation take place, caused by spontaneous development of coarse grassland, shrubs, woodland or forest (Table 12.3a). Fires occur particularly in the vicinity of settlements. After a fire such areas are prone to erosion. Another example is the abandonment of agricultural land on terraces in Southern Europe. After abandonment these terraces are not kept in good condition and fall into disrepair, which is followed by erosion. A second reason for destruction of the terraces is that after abandonment they get colonised by trees. The roots of larger trees damage the terraces, trees fall and the destruction of the terraces is complete, followed by erosion (Table 12.3b). These processes of degradation can cause large landslides, threatening infrastructure and villages lower down. Where no fires occur and soil degradation is not a serious problem, grassland, shrub or woodland vegetation may develop (Table 12.3c). This development has a positive impact on the stabilisation of the dynamic properties of soil by increasing organic matter, in addition to creating vegetation cover that may prevent erosion (Table 12.3c). A special case of the semi-arid regions in Southern and in Central-East Europe (for example Hungary and Bulgaria) are the salt-affected soils. Water management and irrigation measures cause the accumulation of different salts in the topsoil, making further agriculture difficult to impossible. After abandonment these soils slowly develop salt-tolerant vegetation (Table 12.3d). Forest will not develop. The salt-affected soils remain, as the salination process can be considered as an irreversible process on abandoned land.

In the humid and semi-humid regions abandoned arable land will be colonised fast by weeds and soon grassy species will be dominant. Over time

Table 12.3 Vegetation and soil development on abandoned agricultural land as related to disturbance for the main climatic regions

Climate regions	Abandoned agricultural land			References
	Vegetation development	Disturbance	Soil development	
(a) Semi-arid	Forest Coarse grassland Shrubs Open woodland	Fire	Erosion, desertification Nutrient depletion	Bautista et al. (2004); Cerdà (2006); Pardini et al. (2003).
(b) Semi-arid	Trees on terraces	Destruction of terraces	Erosion, landslides	Bautista et al. (2004); Cammeraat et al. (2005).
(c) Semi-arid	Forest Grassland Shrubs Open woodland	No disturbance	Increase soil stabilisation, increase organic matter	Bautista et al. (2004); Rodriguez-Rojo and Sanchez-Mata (2004).
(d) Semi-arid	Salt-tolerant species	No disturbance	Salt-affected soils	Bautista et al. (2004).
(e) Humid Semi-humid	Grassland/shrubs /Forest weeds/ shrubs forest Wetlands	No disturbance	Soil acidification Increase organic matter Increase soil biodiversity Stabilisation soil fertility Nutrient leaching on fertile soils	Bautista et al. (2004); Prach et al. (2000); Hansson and Fogelfors (1998); Bratli et al. (2006).
(f) Humid Semi-humid	Trees on mountain pastures on slopes; Central Europe	Fallen trees	Erosion, landslides	Bautista et al. (2004); MacDonald et al. (2000).

woody species such as shrubs and trees will establish, and the area turns into a forest (Table 12.3e). The speed of these changes depends strongly on soil fertility, soil moisture and climate conditions. Over-fertilised grasslands with unnaturally aggressively growing plants (mostly grasses) create a resistance against the development of forest, whereas forest development on land which is not over-fertilised and has less-fertile soils takes place in a shorter time (Prach et al., 2000). Abandonment of overfertilised lands can have a positive impact on the stabilisation of soil fertility and the soil biological diversity may increase. On the other hand, soil acidification takes place due to the litter accumulation of trees and the lack of liming which was applied during agricultural use. Beyond forest development, wetlands may develop (Table 12.3e). Abandonment of drained lands causes deterioration of the drainage systems and wetland can develop. A special case is the abandonment of alpine regions in Central Europe. In most cases, abandonment of alpine grazing grounds leads to spontaneous regeneration of the forest and restoration of the alpine timberline. In some cases, however, abandonment of mountain pastures can introduce erosion and landslides on slopes with an unstable subsoil (Bautista et al., 2004). The establishment of trees on such unstable sub-soils enhances erosion and landslides (Table 12.3f).

The examples mentioned here show that ecosystem and soil development varies strongly after abandonment of agriculture, and the consequences of this can affect societal use potential and costs of their maintenance. MacDonald et al. (2000) confirm the strong variation in ecosystem and soil development with a series of case studies in mountain areas across Europe. The results of their study show that the sequential nature, direction and scale of the changes in land cover after abandonment are highly variable and hard to predict due to a variety of local circumstances and influences. According to Pardini et al. (2003), abandoned land should be managed to hamper further soil degradation.

Thus management is needed to improve the suitability of abandoned land for other land uses than agriculture, or for the eventually reintroduction of a certain type of agricultural land use system in future.

SUITABILITY OF LAND FOR AGRICULTURE

We have shown in the above sections that the degradation of land and in particularly of soil is due to intensive management and scale enlargement not adapted to land quality, and also due to abandonment of agricultural land. Soil erosion, desertification, compaction, salination, acidification, decline in organic matter and inadequately drained soils contributed mostly to the degradation in the different situations. When the degradation continued it led to a decline in soil productivity and land was no longer suitable for intensive agricultural systems and might become abandoned or used more extensively. Wet, dry, shallow, infertile, saline, stony and difficult-to-cultivate soils, or soils low in soil organic matter, were less suitable or unsuitable to fulfil the requirements of intensive agriculture. Also terrain with small fields, remote fields and those on steep slopes was less suitable or unsuitable for intensive agriculture, and such land was abandoned. Thus it can be said that land having such soil and terrain properties has restrictions for intensive agriculture, and is more suitable for extensive or traditional agricultural systems, or other types of land use better adapted to the land quality. To avoid a decline in productivity, degradation of land quality, and economic setbacks, land use should be adapted to the suitability of land for a certain type of land use. Suitability flow-charts can help to decide on the choice of a proper land use system. Figure 12.1 shows a simplified suitability flow-chart for agriculture when intensification and scale enlargement takes place.

In Figure 12.1 three categories of land suitability for agriculture are distinguished: suitable land (SA category), less-suitable land (LA category) and unsuitable land (UA category). Suitable land (SA) has highly productive and easily cultivatable soils, accessibility of land is good and scale enlargement is possible by consolidation of fields. The climate and soil productivity are favourable for growing several crops. When intensification takes place and is adapted to the local terrain and soil properties – to the static as well the dynamic soil properties – the land is used in a sustainable way for agriculture (SA1 situation). Lowlands with favourable climates and hydrological conditions fulfil these requirements.

When intensification is not adapted to the local terrain and soil properties it will lead to degradation of soil properties and a decline in soil productivity. In the long term this may lead to land abandonment and marginalisation (SA2 situation). The abandoned land should be given a new function to avoid (further) degradation of soil. In the situation of land which is less suitable for agriculture (LA category in Figure 12.1), the quality of land is less suitable for intensification of agriculture as soil or terrain properties do not allow for highly intensive production systems. As long as intensification practices are adapted to these (LA1 situation), this may still lead to sustainable land use, but at a lower level compared to intensification under favourable circumstances (SA1 category). When intensification is not adapted to the local soil productivity and terrain (LA2 situation) this leads to (further) degradation of soil and productivity and to marginalisation. More extensive systems can be an option. However, on the basis of their general lower viability it is to be expected that extensive farming systems are most vulnerable to marginalisation (Hoogeveen et al., 2004). Exploring other types of land use could then be a better option in these areas. This is illustrated in the Czech Republic where the largest changes in land use take place in LFAs. In LFAs abandonment of arable land takes place and turns into grassland, followed by forest development (Boucníková and Kucera, 2005). Forestry is the new function of land use. In the case of unsuitable land for agriculture (UA category in Figure 12.1) indicated by low-productivity soils, steep slopes, inaccessible fields, unfavourable climatic and hydrological conditions, no agriculture should be practised. Such land is probably more suited for other land use. For instance in Estonia strong acid soils are only suited for forestry (Reintam et al., 2003).

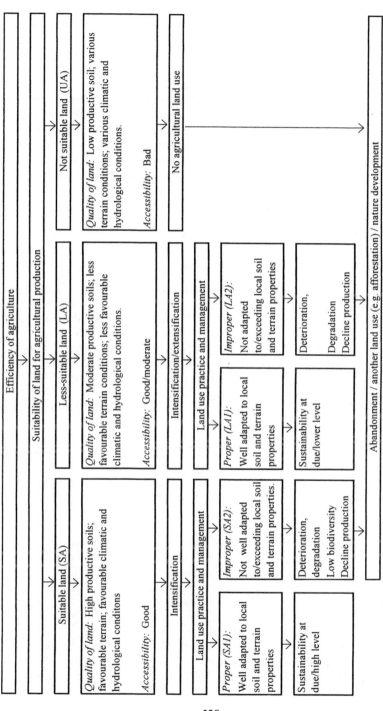

Figure 12.1 Suitability flow diagram for agriculture when intensification and scale enlargement takes place; these are two management decisions feeding the process of marginalisation that eventually cause land abandonment

The suitability flow-chart shows that land quality is important for the decision on the choice of an appropriate land use system. The use of a suitability flow-chart in the decision on the management and type of land use contributes to avoiding economic setbacks and degradation of soil and landscape. At the same time it offers options for other land use. However, to make the information on land quality operational as a tool for the choice of a land use system and management, this information needs to be systemised. This can be done by the development of land classification systems sensitive to the attributes of land quality. Inclusion of land quality in decision schemes on management, measures and land use planning is not only a matter of the field and farm level, but also a matter of larger scales, such as the regional and national level (Bouma, 2002; Boucníková and Fanta, 2005).

LAND CLASSIFICATION AND ITS RELEVANCE FOR POLICY

A broad conceptual framework for a land classification system is given in Table 12.4. In column 1 the suitability of an area for agricultural land use is categorised by using the SA, LA and UA categories of the flow-chart for suitability of agriculture in Figure 12.1. For each suitability category in column 2, main agricultural land use type is given, its goal (column 3), future land use, its function and potential products (column 4) and its environmental impact, separated for proper and improper management (columns 5 and 6). The framework shows that the goals, future land use, its function and products vary for the land use types. It also shows that by proper management each land use type not only has provisions for production but also contributes to the reconfiguration of land utility and function.

Both the suitability flow-chart in Figure 12.1 and the broad conceptual framework for a classification system on land suitability in Table 12.4 can serve as decision tools for the type of land use linked to agriculture or to other land uses. It is possible that when the land is not suitable for agriculture, it may have high value for other land use types such as forestry or nature conservation.

A suitability flow-chart and a land classification system should be part of the sustainable planning of land use not only at the field and farm scale but also at the local, regional and higher scales for developing policy strategies. They may contribute positively to make operational the goals of Agenda 2000 of the CAP and to maintaining the soil functions as defined by the European Commission (2002). In particular they can contribute positively to modulation[1] of the 2003 and 2004 reforms of the CAP. More precisely, this fits in with the concept of integrated land use and management forming a suitable framework for application of the European Community Strategic Guidelines for Rural Development 2007–13 (SEC, 2005), particularly in

Table 12.4   A broad conceptual framework for developing a land classification system

Suitability of area	Main agricultural land use types	Goal	Future land use, its function and products	Possible environmental impact in case of:	
				Proper management	Improper management
SA	Intensive agricultural land use	High production per land unit; economic stability in balance with soil and terrain properties	Sustainable land use; security of production of food and fiber	Maintaining and strengthening landscape structure (Ecol. Networks); preventing land and landscape deterioration and degradation; maintaining and strengthening landscape and biological diversity	Landscape uniformity, loss of linear elements and edges in the landscape; decline biological diversity; soil deterioration (erosion, compaction, acidification, salinisation); water pollution
LA	Extensive land use, intensive but at a lower level as SA category	Harmonisation of land use with natural conditions. Avoiding setbacks in land use.	Alternative methods of sustainable land use (grazing, energy crops, forestry). Strengthening cultural heritage, eco- and agro-tourism	Prevention of abandonment; Improving soil conditions; increasing landscape and biological diversity; restoration of natural biotopes and habitats; strengthening ecological infrastructure of landscape	Abandonment; homogenisation of landscape; soil deterioration: erosion, acidification, saline soils
LA	Continuation of locally traditional land use, HNV farming, extensive farming, extensive organic farming	Conservation of traditional land use forms and methods as cultural heritage and maintaining biodiversity	Sustainable forms and methods of traditional land use; eco- and agro-tourism; handicrafts; local labelled products	Maintaining landscape structure and soil functions; natural and semi-natural biotopes and habitats and biodiversity	Deterioration of landscape as cultural heritage; abandonment
SA, LA	Intensive/extensive organic farming	Production of agricultural commodities of highest quality	Sustainable forms and methods of ecological land use; organic labelled products	Maintaining and/or improving landscape structure and soil functions	
SA, LA, UA	Setting land out of production	Preventing economic setbacks – forthcoming landscape deterioration and/or natural disasters, low yields, high labour costs etc.	Alternative land use: forestry, hunting, recreation, nature and landscape protection	Afforestation, nature development and protection; slope and soil stabilisation	

the first and second axis.[2] Also, a land classification system can contribute at a higher level and in an essential way to the development and maintaining of landscape identity, and may create outstanding conditions to fulfil recommendations of the Pan-European Biological and Landscape Diversity Strategy (Council of Europe, 1996) and the European Landscape Convention (Council of Europe, 2000).

CONCLUSIONS

The relationships of driving forces causing changes in the quality of land are complex. Changes in the quality of land use have their basis grounded in intensification and scale, which if regulated based on the quality of land can serve to conserve and improve land quality and its inherent properties related to productivity, biodiversity and composition. These properties essentially ought to be more lucidly included in local and regional management schemes and policy measures. Suitability flow-charts and land classification systems may help to facilitate this inclusion.

NOTE

1. The mechanism by which EU farm spending is transferred from market-related support payments to rural development policy measures (from pillar 1 to pillar 2 of the European Agricultural Guidance and Guarantee Fund, EAGGF).
2. Axis 1: contribution to a strong and dynamic European agri-food sector by focusing on the priorities of knowledge transfer and innovation in the food chain and priority sectors for investment in physical and human capital. Axis 2: protecting and enhancing the EU's natural resources and landscapes in rural areas. The three EU-level priority areas are: biodiversity and preservation of high nature value farming and forestry systems; water; and climate change.

REFERENCES

Arnalds, A. (2005), 'Approaches to land care: a century of soil conservation in Iceland', *Land Degradation Development*, 16, 113–25.

Arriaga, F.J. and B. Lowery (2003), 'Corn production on an eroded soil: effects of total rainfall and soil water storage', *Soil and Tillage Research*, 71, 87–93.

Bakker, M.M., G. Govers and M.D.A. Rounsevell (2004), 'The crop productivity–erosion relationship: an analysis based on experimental work', *Catena*, 57, 55–76.

Baldock, D., J. Dwyer and M. Sumpsi Vinas (2002), 'Environmental integration and the CAP. A report to the European Commission', DG

Agriculture, Institute for European Environmental Policy (IEEP September 2002).

Ball, B.C., D.J. Campbell, J.T. Douglas, J.K. Henshall and M.F. O'Sullivan (1997), 'Soil structural quality, compaction and land management', *European Journal of Soil Science*, 48, 593–601.

Bautista, S., A. Martinez Vilela, A. Arnoldussen, P. Bazzoffi, H. Böken, D. De la Rosa, J. Gettermann, P.J.G. Loj, J.M. Solera, K. Mollenhauer, T. Olmeda-Hodge, J.M.O. Fernández-Llebrez, M. Poitrnaud, P. Redfern, B. Rydall, J. Sánchez Diaz, P. Strauss, S.P. Theocharopoulus, L. Vandekerckhove and A. Zúquete (2004), 'Measures to combat soil erosion', in L. Van-Camp, B. Bujarrabal, A.R. Gentile, R.J.A. Jones, L. Montanarella, C. Olazabal and S.-K. Selvaradjou (eds), *Erosion*, Reports of the technical working groups, established under the thematic strategy for soil protection, Volume II, European Commission – Directorate-General, Joint Research Centre, European Environment Agency, EUR 21319 EN/2, pp. 199–214.

Bork, H.-R., H. Bork, C. Dalckow, B. Faust, H.-P. Piorr, and Th. Schatz (1998), *Landschaftsentwicklung in Mitteleuropa*, Gotha and Stuttgart: Klett-Perthes.

Boucníková, E. and J. Fanta (2005), 'Landscape planning as a tool to govern landscape development', Conference proceedings Tvář naší země, Vol. 1, pp. 41–6 (in Czech).

Boucníková, E. and T. Kucera (2005), 'How natural and cultural aspects influence land cover structure in Czech Republic', *Ekológia*, 24, 69–82.

Bouma, J. (2002), 'Land quality indicators of sustainable land management across scales', *Agriculture, Ecosystems and Environment*, 88, 129–36.

Bratli, H., T. Økland, R.H. Økland, W. Dramstad, R. Elven, G. Engan, W. Fjellstad, E. Heegaard, O. Pedersen and H. Solstad (2006), 'Patterns of variation in vascular plant species richness and composition in SE Norwegian agricultural landscapes', *Agriculture, Ecosystems and Environment*, 114, 270–86.

Burt, T.P. (2001), 'Integrated management of sensitive catchment systems', *Catena*, 42, 275–90.

Cammeraat, E., R. van Beek and A. Kooijman (2005), 'Vegetation succession and its consequences for slope stability in SE Spain', *Plant and Soil*, 278, 135–47.

Carter, M.R. (2001), 'Organic matter and sustainability', in R.M. Rees, B.C. Ball, C.D. Campbell and C.A. Watson (2001), *Sustainable Management of Soil Organic Matter*, Wallingford: CABI Publishing, pp. 9–22.

Cerdà, A. (2006), 'Forest fires are not so bad: a case study in Spain', in A. Imeson, O. Arnalds, L. Montanarella, A. Arnoldussen, S. Van Asselen, L. Dorren, M. Curfs and D. De la Rosa (eds), *Soil Conservation and Protection in Europe: The Way Ahead*, EUR 22187 EN, Heiloo, the Netherlands, pp. 59–62.

Commission staff working document (2005), 'Annex to the: Proposal for a Council Decision on Community strategic guidelines for Rural Development' (Update to Impact Assessment Report (SEC(2004) 931). COM (2005)304 final.

Council of Europe (1996), 'Pan-European Biological and Landscape Diversity Strategy', Sofia – Strassbourg.

Council of Europe (2000), European Landscape Convention. Florence – Strassbourg, http://conventions.coe.int.

De la Rosa, D., J.A. Moreno, F. Mayol and T. Bonsón (2000), 'The assessment of soil erosion vulnerability in Western Europe and potential impact on crop productivity due to loss of soil depth using the ImpelERO model', *Agriculture, Ecosystems and Environment*, 81, 179–90.

Dumanski, J. and C. Pieri (2000), 'Land quality indicators: research plan', *Agriculture, Ecosystems and Environment*, 81, 93–102.

Duram, L.A., J. Bathgate and Chr. Ray (2004), 'A local example of land use change: Southern Illinois – 1807, 1938, and 1993', *Professional Geographer*, 56 (1), 127–40.

EEA (European Environmental Agency) (2001), 'Surface area of sites with dry or mesic grassland habitat types proposed for nature protection under the flora, fauna and Habitats directive (pSCIS)', http://themes.eea.int/Specific_media/nature/indicators/grasslands/protectio n/yir01bio.pdf.

European Commission (2002), 'Communication from the Commission to the Council, the European Parliament, the Economic and Social Committee and the Committee of the Regions. Towards a Thematic Strategy for Soil Protection', COM (2002) 179.

Firbank, L.G., S.M. Smart, J. Crabb, C.N.R. Critchley, J.W. Fowbert, R.J. Fuller, P. Gladders, D.B. Green, I. Henderson and M.O. Hill (2003), 'Agronomic and ecological costs and benefits of set-aside in England', *Agriculture, Ecosystems and Environment*, 95, 73–85.

Flinn, K.M., M. Vellend and P.L. Marks (2005), 'Environmental causes and consequences of forest clearance and agricultural abandonment in central New York, USA', *Journal of Biogeography*, 32, 439–52.

Gunreben, M. (2005), 'Dealing with soil threats in Lower Saxony, Germany', *Land Degradation and Development*, 16, 547–50.

Hajkowicz, S. and M. Young (2005), 'Costing yield loss from acidity, sodicity and dryland salinity to Australian agriculture', *Land Degradation and Development*, 16, 417–33.

Hansson, M. and H. Fogelfors (1998), 'Management of permanent set-aside on arable land in Sweden', *Journal of Applied Ecology*, 35, 758–71.

Håkansson, I. and V.W. Medvedev (1995), 'Protection of soils from mechanical overloading by establishing limits for stresses caused by heavy vehicles', *Soil and Tillage Research*, 35, 85–97.

Hoogeveen, Y., J.-E. Petersen, K. Balazs and I. Higuero (2004), 'High nature value farmland: characteristics, trends and policy challenges', Copenhagen: European Environment Agency, EEA report 1/2004.

Kosmas, C., St Gerontidis and M. Marathianou (2000), 'The effect of land use change on soils and vegetation over various lithological formations on Lesvos (Greece)', *Catena*, 40, 51–68.

Küster, H. (1999), *Geschichte der Landschaft in Europa: von der Eiszeit bis zur Gegenwart*, München: Beck.

Lægreid, M., O.C. Bøckman and O. Kaarstad (1999), *Agriculture, Fertilizers and the Environment*, Wallingford, UK and New York, USA: CABI Publishing, CAB International.

MacDonald, D., J.R. Crabtree, G. Wiesinger, T. Dax, N. Stamou, P. Fleury, J. Gutierrez Lazpita and A. Gibon (2000), 'Agricultural abandonment in mountain areas of Europe: Environmental consequences and policy response', *Journal of Environmental Management*, 59, 47–69.

Meier, K., V. Kuusemets and Ü. Mander (2005), 'Socio-economic and land-use changes in the Pedja River catchment area, Estonia', in C.A. Brebbia and J.S. Antunes do Carmo (eds), *River Basin Management III*, WIT Transactions on Ecology and the Environment, Vol. 83, WIT Press, UK.

Morari, F., E. Lugato, A. Berti and L. Giardini (2006), 'Long-term effects of recommended management practices on soil carbon changes and sequestration in north-eastern Italy', *Soil Use and Management*, 22, 71–81.

Mottet, A., S. Ladet, N. Coque and A. Gibon (2006), 'Agricultural land use change and its drivers in mountain landscapes: a case study in the Pyrenees', *Agriculture, Ecosystems and Environment*, 114, 296–310.

Nertsen, N.K., O. Puschmann, J. Hofsten, A. Elgersma, G. Stokstad and R. Gudem (1999), 'The importance of Norwegian Agriculture for the Cultural Landscape – A subproject under the Ministry of Agriculture's evaluation programme on multifunctional agriculture', NILF-notat 1999:11.

Pardini, G., M. Gispert and G. Dunjó (2003), 'Runoff erosion and nutrient depletion in five Mediterranean soils of NE Spain under different land use', *Science of the Total Environment*, 309, 213–24.

Peterson, U. and R. Aunap (1998), 'Changes in agricultural land use in Estonia in the 1990s detected with multitemporal Landsat MSS imagery', *Landscape and Urban Planning*, 41, 193–201.

Prach, K., I. Bufková, F. Zemek, M. Heřman and Z. Mašková (2000), 'Grassland vegetation in the former military area Dolbrá Voda, the Šumava National Park', *Silva Gabreta*, 5, 101–12.

Puigdefabregas, J. (1998), 'Implications of regional scale policies on land condition and land degradation in the Mediterranean basin', in Global Change and Terrestrial Ecosystems, The Earth Changing Land, GCTE-LUCC, Open Science Conference on Global Change, GCTE, Barcelona.

Pulleman, M., A. Jongmans, J. Marinissen and J. Bouma (2003), 'Effects of organic versus conventional arable farming on soil structure and organic matter dynamics in a marine loam in the Netherlands', *Soil Use and Management*, 19, 157–65.

Reidsma, P., T. Tekelenburg, M. van den Berg and R. Alkemade (2006), 'Impacts of land-use change on biodiversity: an assessment of agricultural biodiversity in the European Union', *Agriculture, Ecosystems and Environment*, 114, 86–102.

Reintam, L., A. Kull, H. Palang and I. Rooma (2003), 'Large-scale soil maps and a supplementary database for land use planning in Estonia', *Journal of Plant Nutrition and Soil Science*, 166, 225–31.

Rodriguez-Rojo, M.P. and D. Sanchez-Mata (2004), 'Mediterranean hay meadow communities: diversity and dynamics in mountain areas throughout the Iberian Central Range (Spain)', *Biodiversity and Conservation*, 13, 2361–80.

SEC (2005), 'Proposal for a Council Decision on Community strategic guidelines for Rural Development, Programming period 2007–2013', presented by the Commission, SEC(2005) 914, Commission of the European Communities.

Spek, Th. (2004), 'The Drenthe *esdorpen* landscape: a historical-geographical study', PhD Thesis, Wageningen University (in Dutch).

Stockfisch, N., T. Forstreuter and W. Ehlers (1999), 'Ploughing effects on soil organic matter after twenty years of conservation tillage in Lower Saxony, Germany', *Soil and Tillage Research*, 52, 91–101.

Watson, C.A., D. Atkinson, P. Gosling, L.R. Jackson and F.W. Rayns (2002), 'Managing soil fertility in organic farming systems', *Soil Use and Management*, 18, 239–47.

Wiebe, K. (2003), 'Linking land quality, agricultural productivity, and food security', Agricultural Economic Report Number 823, USDA/Economic Research Service.

Young, J., A. Watt, P. Nowicki, D. Alard, J. Clitherow, K. Henle, R. Johnson, E. Lacxko, D. McCracken, S. Matouch, J. Niemela and C. Richards (2005), 'Towards sustainable land use: identifying and managing the conflicts between human activities and biodiversity conservation in Europe', *Biodiversity and Conservation*, 14, 1641–61.

13. Emerging perspectives on changing land management practices

Floor Brouwer, Teunis van Rheenen and Shivcharn S. Dhillion

CURRENT PERSPECTIVES

This volume shows that marginalisation is not limited by geographical location. Although the context and the manner by which it is manifested is location-specific, the basic drivers behind the processes are universal. While the analysis largely draws on cases within Europe, the complementary analyses undertaken in other parts of the world display remarkably similar processes. This reiterates the desire for tools that are generic and sufficiently flexible to accommodate local and regional characteristics.

Our Current Understanding of Marginalisation

According to a dictionary of sociology, marginalisation is a process by which a group or individual is denied access to important positions and symbols of economic, religious or political power within any society. In the specific context of land management practices, marginalisation of land is a process of changing land management practices, driven by a combination of social, economic, political and environmental factors by which the use of land for the main land-dependent activities (agriculture, forestry, housing, tourism, local mining) ceases to be viable under an existing socio-economic structure. Marginalisation can be understood as a multidimensional process encompassing not only land abandonment, environmental degradation and economic decay, but also change in social and cultural patterns. The concept of marginalisation, related both to problems of viability of agricultural activity in less-favourable areas and to the decay of rural communities, was debated in the European context in the late 1980s and the beginning of the 1990s. The process of marginalisation thus clearly applies to agriculture when in certain areas farming ceases to be viable under an existing land use, socio-economic and policy frame, and no other agricultural options are available. Marginalisation goes beyond the European context and applies to that of developing nations where policy reformations, land use dynamics, agricultural practices and cropping choices, and climate change are consistently taking on new dimensions. There are different stages in the process of marginalisation, including land abandonment and the cessation of the management of land.

The process of marginalisation may end at land abandonment, occurring at various scales ranging from the parcel (plot) or farm to regional levels, and it can have compounded impacts on the landscape. Marginalisation is mainly observed in remote areas, either because of biophysical constraints, but even more importantly due to demographic and socio-economic factors limiting viability of land management practices. Desertification, for example, and related productivity potential, have long been key factors in limiting viable management of the land. More recently, social processes (like depopulation), economic conditions, and national and regional land use policies have gained importance in explaining the configuration of marginalisation.

Several projections on future land use patterns include high rates of abandonment of agricultural land, even in countries with modern and competitive agriculture, since under the foreseen conditions it would not be economically rational to continue farming on land with less-favourable conditions or locations (Baldock et al., 1996; Bethe and Bolsius, 1995; Cabanel and Ambroise, 1990; Reenberg and Pinto-Correia, 1993). However, the empirical evidence on land abandonment remains poor. This may in part be due to the fact that classification of land use varies across cultures, regions and nations, often with vague boundaries between extensive and intensive land use. Also, the frameworks that link marginalisation with land abandonment are scarce and need further development, especially as there is a growing acceptance for the recognition of marginalisation as a process among many governments and communities which for a long time ignored its reality.

Factors Driving Marginalisation

Factors that drive marginalisation of agriculture include the ageing of the farming population, decline of the importance of agriculture in the rural economy, and the decreasing number of farms. Urbanisation also is an important factor that puts pressure on marginalisation of farming. People move to the countryside for a variety of reasons; regardless of the reason, the shift does change the rural countryside. Land values increase as the demand increases. When evaluating processes of change it is important to consider more than the net impact of the changes. In other words, the impact of change is not just the total of the benefits and costs but also the distribution of those benefits and costs.

Diversity in land management practices increases over time, with high demand of farmland for non-farm use. The majority of farmers are small farmers who do not rely on farming for the bulk of their income. Agricultural production on about a quarter of the holdings in the USA is worth less than US $1000 per annum, representing only 0.5 per cent of total sales and 9 per cent of the farmland. Fifteen per cent of the farms report almost 90 per cent of total sales in agricultural produce. The trend towards a dual-faceted agriculture – with intensification and extensification processes – appears to be accelerating. The number of holdings remains to increase in many developing countries, reflecting differences in stages of economic development. A decline in the number of agricultural holdings can be observed in certain regions of China for the first time, as in countries like Malaysia and Singapore where holdings are becoming larger and land use is changing and being rapidly redefined (both nationally and regionally, for example ASEAN – Association of South East Asian Nations) to meet social, economic and ecological demands. This also illustrates a transition in agricultural production structures due to high economic growth with large labour demand outside agriculture.

Marginalisation in Remote Areas

Some regions at the edge of the European Union (EU) face high unemployment rates and a high proportion of economic inactivity. The level of economic development is insufficient to support the shift away from farming towards other economic activities. Hence, agricultural and rural support programmes offer an important basis to improve the viability of such rural regions. However, social capital is vital to achieve the further development and diversification of regional economies and the provision of services to society. A shift of political and economic boundaries – like the enlargement of the EU – causes a change in economic activities, and hence in the process of marginalisation. Multifunctionality might benefit from new economic activities. Under these circumstances, conventional agriculture is losing its ability to lead the way in maintaining the viability of rural areas. The interweaving of complementary activities with agriculture is paving the road to multifunctionality.

Marginalisation at the Edge of the Urban Fringe

Marginalisation might be limited in regions that also face economic decline and are at a short distance to main urban centres. This could take place in regions where alternative labour conditions are available. Some mountain regions in Austria, for example, face economic decline due to limited possibilities for tourism. In such cases, the available infrastructure and other facilities (for example roads, public transport and shops) will determine the extent to and the rate at which marginalisation occurs.

The underlying concepts are no longer based so much on considerations for preservation, but are increasingly inspired by higher valuation and outside demand for the unique resources of mountain areas. A redefining of land, its uses and potential beyond agriculture or forestry, may facilitate the sustainability of the new, growing demands. An integrated approach is necessary to take account of the multi-sectoral aspects of resource use and socio-economic development in mountain areas.

Marginalisation in Regions with a Concentration of Agricultural Production

The share of agriculture in regional economies generally shows a declining trend. Agriculture however remains a major activity in some parts of Europe. For example, agriculture accounts for more than 40 per cent of total employment in parts of Spain, and as such is a crucial factor for the proper management of the land and a main determinant of marginalisation.

In parts of Spain, the density of population is just around ten inhabitants per square kilometre. This level is identified as the minimum threshold to sustain a social service network at the county level making agricultural activities important drivers for marginalisation.

Cross-Compliance to Prevent Land Abandonment

The 2003 CAP reform introduced the requirement for farmers in receipt of direct payments to comply with certain standards in relation to the environment, and to public, animal and plant health and animal welfare, or face reductions in, or withdrawal of, those payments. This system of attaching conditions or requirements to farm payments is known as cross-compliance. Some of these standards are based on pre-existing European legislation and are called Statutory Management Requirements (SMRs). Other standards are aimed at ensuring Good Agricultural and Environmental Conditions (GAECs) and reflect the issue of soil management and the minimum maintenance of agricultural land. GAEC standards related to land abandonment seem to be sufficiently clear to farmers in the United Kingdom (Moravec and Zemeckis, 2007). Standards are introduced to keep the land in such a state that it can be returned to good agricultural condition. Emphasis is given to maintaining the productive capacity of land, and land abandonment is not aimed for in a direct manner.

Multifunctionality and the Science–Policy Debate

Multifunctionality has been widely debated during the 1990s, in the political context of reforming agricultural policy. The public debate on multifunctionality includes trade-related concerns, as well as rural and agricultural development models. Many concepts are available to enhance the understanding of multifunctionality, often linked to scientific disciplines, resulting in Concept Oriented Research Clusters (CORCs). However, an operational tool for multifunctionality is still lacking. The boundaries between sectoral activities (for example farming, forestry and tourism) remain vague and subject to interpretation. The CORC is promising, yet will need a more holistic perspective.

Simultaneously, research has been instrumental and provided an insight of such transformations through the mobilisation of various disciplines (for example agronomy, economics, sociology). Yet a shortcoming is that it has remained largely case-study based. Little information is, for example, provided on the socio-economic impact of alternative and diversified farm activities in terms of their contribution to income and rural employment.

Multifunctionality could be a concept to understand sustainable development better. Multifunctionality strengthens our body of knowledge on sustainable development, making the linkages between the different components transparent. For example, farmers producing food and meanwhile also maintaining the landscape contribute to the economic, social and ecological dimensions of sustainable development.

Multifunctional agriculture can help forge the bond between agriculture and the surrounding rural communities as a desirable place to live. Only time will tell if this new approach helps alleviate the problems we are seeing in agriculture and rural communities. The concept of multifunctional agriculture is little understood, explored or used in the developing world, and its applicability in both conceptual and practical senses requires serious consideration.

NEW PERSPECTIVES ON LAND MANAGEMENT PRACTICES

Pluriactivity is vital for long-term viability of land management practices in remote areas. It refers to the extent to which farmers have complemented their income by gainful activities outside agriculture, that is, every activity other than activity relating to farm work, carried out for remuneration (salary, wages, profits or other payment, including payment in kind, according to the service rendered). Pluriactivity will facilitate the existence of multifunctional land use. However, some land management practices may already supply a range of different complementary functions (for example food production with maintenance of open landscapes), serving economic, cultural and environmental needs.

Agriculture in Finland and Norway is dependent on farm support programmes. There are two possible strategies to cope with marginalisation in Norway and Finland: one is increasing farm size or changing to more productive and specialised crops; the other option for farms to survive is pluriactivity. Such activities beyond farming require a demand by society (for example a society that is prepared to pay for open landscapes). A shift from full-time to part-time farming indicates lower returns to investments in the agricultural sector, forcing the farmer communities to seek additional income from other sources. The integration of the farming community into the rest of society is a vital strategy to cope with marginalisation and to broaden the economic basis for viable land management practices. However, where the farming communities make up most of the population, societal interventions need to be significant.

The availability of off-farm jobs is crucial for combating marginalisation. However in sparsely populated areas with long distances to centres, finding an off-farm job may be a problem.

Response by Consumers to Reduce the Vulnerability to Marginalisation

Infrastructure and markets need to be in place to achieve high nature value farming systems. Important factors are consumers to buy the food, and civil society to express willingness to pay for the delivery of services by farmers. As is the case in the EU, in Japan there is a growing, high level of interplay between agricultural production and nature conservation.

Understanding Part-time Farming in the Context of Marginalisation

The extent of part-time farming may indicate the occurrence of holdings with insufficient work to fulfil the amount of work required for an annual work unit (AWU). This reflects the potential need for additional activities in order to maintain a viable agriculture. Regions might be susceptible to the occurrence of marginalisation in cases where they have a high share of farms without sufficient other gainful activities that also have insufficient activities to meet the requirements to fulfil the standards for AWUs. Part-time farming is measured as farm holders working less than one AWU on the farm (Figure 13.1). The share of part-time farming exceeds 80 per cent in Greece, Spain, Italy and several of the ten member states who entered the EU in 2004 (Estonia, Latvia, Lithuania, Hungary, Malta, Slovakia and Slovenia).

The Need for Understanding Social Cohesion and Social Capital to Cope with Marginalisation

Since our analysis points out clearly the importance of social capital in rural development dynamics, we suggest that social capital should be better recognised by policy-makers as a key factor in the rural development process, hampering (when weak) or helping (when strong and well rooted) the implementation of rural development policies and specifically those policy measures aimed at counteracting marginalisation processes and, to some extent, land abandonment.

Challenges of Climate Change: Needs for Adaptive and Risk Management Measures

There is no doubt that the Intergovernmental Panel on Climate Change (IPCC) has put out extensive reports concerning the sensitivity, adaptive capacity and vulnerability of natural and human systems to climate change. Although not all the assumptions about (future) impacts are reported with the highest confidence level, there are certainly clear impacts and predictions that we consider relevant here. The impact predictions point to crop productivity (differential based on the type of crop) to increase slightly in mid to high latitudes where increases of temperature can be in the range of 1–3°C: but this is expected to have mixed effects due to the parallel projected increase in the frequency of winter/spring floods, ground instability and runoff. At lower latitudes (where agriculture has its base globally), especially in dry (e.g., including many parts of southern Europe) and tropical regions, crop productivity is projected to decrease for even small local temperature rises (1–2°C). Drought, higher temperatures and flood frequency increases are expected to affect crop production negatively by reducing water availability (higher water stress). The most vulnerable in this case are those in subsistence sectors at low latitudes. Several related sectors will be affected, including hydropower, tourism (both summer and winter), water supply works, and infrastructure builders (due to level ground instability). Marginalisation can occur regardless of location and arable lands at the fringes may experience abandonment at a higher rate and forest takeover, as already occurring at present.

Climate change can thus act as a driving force for increased food prices (as is experienced during drought and flood periods the world over) affecting particularly smallholdings and the poor). Certainly based on the projections some areas (northern Europe and higher latitudes areas) will have an increase in productivity, and longer growing seasons coupled with crop choice changes, and this may work to reduce the risk of marginalisation of arable land if the negative more physical impacts do not outweigh the benefits. The most vulnerable projected areas are the tropics and lower latitude areas, changes affecting grain production particularly, where consequences can have severe impacts on regional and local food security and social-economic conditions. For example, Tubiello and Fisher (2007) estimate that while global cereal production would not decrease significantly by 2080 due to climate change, the regional effects would be dramatic, with declines of over 22 per cent in South Asia (see also discussions in EEA 2005; Riahi et al., 2006; López-Moreno et al., 2008; IPCC reports at www.ipcc.ch). As endowments of land, labour and capital change globally, so may water endowments and distribution among land use classes, making national agricultural choices, driven in part by global economic trade, more susceptible to vulnerability under climate changes. Here multifunctional approaches have to be flexible and adaptive to combat predicted (and yet unpredicted locally) climate changes, and proactive climate change risk management and mitigation plans in the land use context have to be enacted to balance economic, social and environmental functions. Marginalisation may yet have different trajectories based on complex interactions of climate change and socio-economic policy responses, both nationally and internationally.

Lessons for Policy and Research

A new perspective on marginalisation and abandonment might be essential when considering climate change. Differential responses to climate change locally and regionally will have to be integrated into strategies for combating marginalisation, configurations of which may shift geographically and have detrimental localised impacts without altering significantly in magnitude globally.

Political pressure on agricultural budgets remains large, and opposition to large-scale farm payment programmes has grown steadily. A reduction in agricultural support programmes puts pressure on farming in remote areas. A reduction in agricultural budgets may therefore also increase the risk of marginalisation of agricultural land. The integration of environmental and rural concerns in agricultural policy is a main issue in political debate. Negotiations to liberalise trade in the context of the World Trade Organization (WTO) exerted pressure for the decoupling of farm support programmes from production. Without additional measures and instruments, such programmes may increase the risk of marginalisation. This is acknowledged in the 2003 reform of the Common Agricultural Policy (CAP), with farmers that need to implement measures for GAECs.

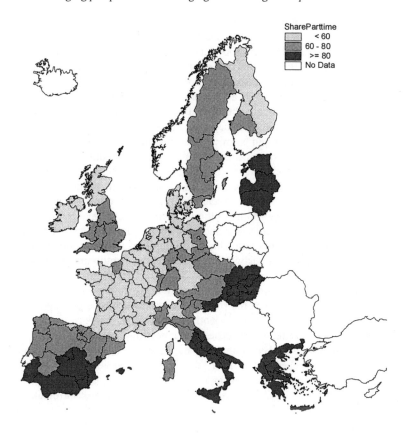

Note: Hamburg, Bremen, Berlin (Germany), Austria and Portugal in 2000.

Source: Eurostat (FSS/Eurofarm).

Figure 13.1 Share of part-time farming in agriculture (holders working less than 100% of an AWU) in 2003 (%)

Marginalisation of agricultural land and its eventual abandonment both have a negative connotation and imply degradation. However, this is not necessarily the case in practice. There are numerous examples where re-configuration of former agricultural land has given rise to more productive and economically viable functions (for example nature parks, tourist attractions). To make this sustainable requires a demand for the search for such alternative functions of land.

Combating marginalisation in remote and sparsely populated regions may be costly, and measures to control marginalisation of agricultural land need to be compared to other land management practices that sustain functions supported by agriculture and demanded by society. There is an urgent need to quantify the costs of the supply by farmers of functions other than food, and to compare them to the costs of other land management strategies.

There is a need to understand urban–rural linkages better. Cities benefit from the functions supplied by agriculture in rural areas. This notion is barely acknowledged in the linkages between cities and rural areas. Revisiting and rethinking of the traditional concepts of spatial planning, rural development measures, fiscal regimes and labour market instruments are required.

A framework to determine the linkages for marginalisation has not yet been developed. In this context, quantitative research on marginalisation and the incorporation of these results in land management processes will be a crucial element in helping us to understand marginalisation processes better. Also, there is a great need to be able to determine at a very early stage areas that are at risk of becoming marginalised. Risks need to be considered in a broader perspective, taking into consideration severity, likelihood and timing.

Marginalisation is an early warning for future abandonment of land. To avoid this, we need to understand the signals. This volume has described many of the possible signals which should make one aware of potentially undesirable developments in the rural areas. Indicators have proven to offer a scientifically robust approach to identify the diversity of signals that depend on the socio-economic, demographic and political context of rural areas. However, modelling tools dealing with marginalisation and multifunctionality of land in a comprehensive manner still remain to be developed and tested. Climate scenarios need due consideration.

Today, the future of the rural areas is at stake. The core question until now has been: What policies are required to maintain agriculture in the rural areas? This is however not totally inclusive and is inadequate to address marginalisation. In addition, we need to explore more rigorously which alternative and adaptive functions land can fulfil to meet the challenges of tomorrow.

REFERENCES

Baldock, D., G. Beaufoy, F. Brouwer and F. Godeschalk (1996), *Farming at the Margins: Abandonment or Redeployment of Agricultural Land in Europe*, London and The Hague: Institute for European Environmental Policy (IEEP)/Agricultural Economics Research Institute (LEI-DLO).

Bethe, F. and E. Bolsius (eds) (1995), *Marginalisation of Agricultural Land in Europe: Essay and Country Studies*, The Hague: National Spatial Planning Agency.

Cabanel, J. and R. Ambroise (1990), 'La France part en friche ... et alors?' *Metropolis*, 87, 4–11.

EEA (2005) Vulnerability and adaptation to climate change in Europe. Technical Report no. 7/2005. (www.reports.eea.europa.eu).

López-Moreno, J.I., M. Beniston and J.M. Garcia-Ruiz (2008), 'Environmental change and water management in the Pyrenees: facts and future perspectives for Mediterranean mountains', *Global and Planatery Change*, 61 (3–4), 300–312.

Moravec, J. and R. Zemeckis (2007), 'Cross-compliance and land abandonment', A research paper of the Cross-Compliance Network (Contract of the European Community's Sixth Framework Programme, SSPE-CT-2005-022727), Deliverable D17 of the Cross-Compliance Network.

Reenberg, A. and T. Pinto-Correia (1993), 'Rural landscapes and marginalization: can general concepts, models and analytical scales be applied throughout Europe?' Leipzig-Halle, Germany: EUROMAB, 3, 3–18.

Riahi, K., A. Grüber and N. Nakicenovic (2007), 'Scenarios of long-term socio-economic and environmental development under climate stabilization', *Technological Forecasting and Social Change*, 74 (7), 887–935.

Tubiello, F.N. and G. Fisher (2007), 'Reducing climate change impacts on agriculture: Global and regional effects of mitigation, 2000–2080', *Technological Forecasting and Social Change*, 74 (7), 1030–56.

Index